My Adventures With Your

The color dust jacket is by Las Vegas artist Roy E. Purcell. Entitled, "The Promoter," the etching depicts a mining camp entrepreneur in a central Nevada mining camp during the height of speculation in 1907.

Copyright © 1986 by Stanley W. Paher
All rights reserved

ISBN 0-913814-75-X

PRINTED IN THE UNITED STATES OF AMERICA

Published by Stanley W. Paher

MY ADVENTURES WITH YOUR

by George Graham Rice

Introduction
by
Hugh
Shamberger

Nevada Publications
Post Office Box 15444
Las Vegas, Nevada 89114

Table of Contents

	Introduction by Hugh Shamberger	9
	Foreword by George Graham Rice	25
1.	The Rise and Fall of Maxim and Gay	29
2.	Mining Finance at Goldfield	53
3.	A Saturnalia of Speculation	97
4.	The Greenwater Fiasco	135
5.	The Great Goldfield Smash	147
6.	Nipissing & Goldfield Consolidated	175
7.	Birth of Rawhide	211
8.	Publicity and the Public's Money	221
9.	The Wall Street Game	257
10.	B. H. Scheftels & Company	273
11.	A Fight to the Death	287
12.	The Lesson of It All	325
	Index	327

During the height of speculation and investment in 1907, Goldfield was Nevada's largest city.

George Graham Rice, Nevada promoter extraordinaire

Introduction

by Hugh Shamberger

THE NAME George Graham Rice, although a pseudonym, appears frequently in historical accounts of Nevada's 20th century mining camps, as well as in shady stock promotional circles all across the nation. During his lifetime Rice saw several millions of dollars pass through his hands, but very seldom was he able to keep any of it. He was a heavy gambler, but also spent money lavishly and constantly had a stable of highly talented lawyers on hand—both for his promotional operations as well as for his many court trials.

Born Jacob Simon Herzig, this brilliantly talented confidence man was first convicted of larceny in New York in 1890. Since he was 19 at the time, his sentence consisted of a two-year stretch in an Elmira reformatory. A second conviction in 1893, this time for forgery, led to a six year sentence at Sing Sing, although he was released after serving four years.

Following this second term of imprisonment, Herzig underwent a major change in his approach to life. Adopting the alias of George Graham Rice, he established a horse race tipping service in New York City under the name of Maxim & Gay. In the next two years the business earned more than $1.5 million, but as Rice was a spendthrift as well as an inveterate gambler, he spent the money as fast as it came in.

It was not long before the U.S. Postal Service began to investigate Rice's use of the mail, and the local district attorney's office also became interested in his affairs. Rice quickly discontinued the Maxim & Gay service, only to find himself broke by the spring of 1904. His credit was still good, particularly with race-track book makers, so he became an active bettor. By June 1904, he had amassed $100,000, but by the following month he was once again broke. Finding few new avenues to instant wealth and success in New York, Rice headed west to try his luck in the new mining camps springing up throughout Nevada.

IN THE FALL OF 1904 Rice reached central Nevada with only $150 in his pocket. He started the Goldfield-Tonopah Advertising Agency and within six months had netted $65,000. Rice then organized a news bureau as an adjunct of the advertising agency, sending off "news" articles about the mining booms in Bullfrog, Manhattan and other thriving camps. During these early days in Goldfield Rice made large sums of money writing and submitting advertising material about certain mines to newspapers in California, Salt Lake City and Denver. But as quickly as he was "in the money," he gambled it away.

Among Rice's favorite games of chance was faro, which he often played in the Palace Club, Goldfield's second-largest gambling hall, which was owned by "Larry" Sullivan and Peter Grant. One evening, as Rice was cashing in some $2,500 in winnings, Sullivan asked if he could be cut in on some of Rice's mining deals. Rice replied that if Sullivan would match the $2,500 in front of him then he'd see what could be done. Sullivan quickly put up the money. The next day Rice sent him off to Manhattan to purchase the Jumping Jack mine, and within three weeks they had cleared $250,000.

Shortly thereafter the L. M. Sullivan Trust Company was formed with Sullivan as president and Rice as vice-president and general manager. Sullivan divided his interest with Peter Grant, his gambling house partner, who remained in the background throughout the life of the company, which lasted from late summer of 1905 to January 1907.

Before coming to Goldfield, Sullivan ran a sailors' rooming house in Portland which had developed a reputation as one of the lowest dives in the Pacific Northwest. So many of Sullivan's "guests" woke up aboard tramp steamers bound for the Orient that the proprietor soon became known as "Shanghai Larry." The *Financial World of New York* later stated that "while he is supposed to be a gun fighter, he never shot anyone unless the victim's back was turned . . ."

In contrast, Rice was described as a dapper man standing about five feet ten inches in height, slightly built and with well-groomed black hair. Cultivating a flair for fashionable clothes, he often changed suits four times a day. He was never without a full wallet and a pocketful of cigars, and was generous with both. Rice was also a persuasive speaker, as one of his habits was to learn a new word each day from the dictionary.

With Rice masterminding the operation, the L. M. Sullivan Trust Co. was an instant success. At one time it operated as many as seven properties with a combined payroll of $50,000. These companies, like most of the 200 mining companies in Goldfield, each had a capitalization of one million shares of stock with a par value of $1 per share. In exploiting these companies the stock was put on the market for ten cents a share, after which great publicity and sales effort was used to run up the price and reap the rewards.

Some of the money thus generated was actually used to develop the various mines. In fact, this was the method used by most mining companies, not only in Goldfield but in mining camps throughout the West. In 1910 noted mining engineer Augustus Locke, writing in the *Mining and Scientific Press*, explained how this method was used to finance the Goldfield mines:

> ... There are about 200 companies promoted in Goldfield. Almost invariably they have a capitalization of 1,000,000 shares. Each company sold, on an average, one half of its capital stock. The average price realized is unknown, but likely was 30 cents. The companies which have since had great success are included in this average, but as their original selling price was low, they bring up the average little or none. The money invested then in the initial purchase of the mining securities of Goldfield was something like $30,000,000.

The great difference between the Rice operations and others was the nationwide publicity given to the L. M. Sullivan Trust Company stock by newspapers throughout the nation. Rice also utilized direct mailings sent to people on the company's 65,000-name "sucker" list. Unfortunately, it was this latter practice which once again brought U.S. postal authorities looking into his operations.

Soon the L. M. Sullivan Trust Co. was promoting mines in Goldfield, Manhattan, Bullfrog, and Greenwater (in California), and some of their properties were indeed good bets for investors. The Lou Dillon enjoyed an excellent location close to both the rich Combination and Mohawk claims, although it unfortunately failed to develop as a paying mine. Another property controlled by the Sullivan Trust was the Jumbo Extension, which eventually developed into a rich producer under different ownership. Rice later stated that all the mines he promoted were genuine with the exception of only one —the Bullfrog Rush Company at Bullfrog. In this instance, claimed Rice, the L. M. Sullivan Trust Co. refunded $90,000 to investors.

FINALLY IN JANUARY 1907 the Sullivan Trust Co. failed. It was overextended and operating on credit, having converted $2 million in assets into an estimated $1.2 million in liabilities. With the firm's demise, the market value of all Goldfield stocks plunged, and the general public began to lose confidence in these mines.

According to Rice the decline and fall of Goldfield as a mining camp dated from the hour the Sullivan Trust Company failed. He claimed that "the *Goldfield News,* of nation-wide circulation and up to then unshackled, sought to stem the tide. It published a double-leaded editorial, in full-face type, setting forth that the Sullivan Trust Company had gone down with its flag nailed to its masthead of a declining market and had lost its last dollar supporting its own stocks."

With federal postal authorities looking into his affairs and with the collapse of the L. M. Sullivan Trust Company, Rice felt it expedient to leave Goldfield, which he did late in January 1907. He returned to New York, supposedly with just $200 to his name, but found conditions there not to his liking. "In New York's financial mart," he wrote, "I felt like a minnow in a sea of bass." Within a few weeks he was on his way back to Nevada.

This time Rice selected Reno as his headquarters, but it was the bustling mining camp of Wonder which next attracted his attention. On arriving there he ran across his former partner Larry Sullivan whereupon the two decided to join forces in promoting a Wonder property. Sullivan put up the money to buy the Rich Gulch group of claims, then the new Sullivan & Rice corporation went to work promoting the Rich Gulch Wonder Mining Company with a million-share capitalization.

A high-class directorate was secured with J. F. Dunnaway, vice-president and general manager of the Nevada, California & Oregon Railroad as president; Nevada's Governor John Sparks as vice-president; U. S. Webb, Attorney General of California, the second vice-president; and D. B. Boyd, Washoe County Treasurer, as the firm's treasurer.

But try as he might, the Rich Gulch promotion failed to interest any large number of investors. "The Wonder mining camp boom had died abornin'," lamented Rice. "Investors seemed to have had enough of mining stock speculation for a while."

Rice next decided to help finance a new Reno newspaper venture which would report on mining throughout the state. Merrill A. Teague, recently on the staff of a Goldfield daily market sheet called the *Nevada Mines News Bureau,* was made editor of the fledgling *Nevada Mining News* in early May 1907.

"Mr. Teague is a possessor of a facile pen," wrote Rice. "I was convinced that the *Nevada Mining News* had a cheap editor. When news was scarce he could write more about nothing than any man I ever met before. Incidentally, he could go further without finding a stopping place in a crusade than any man I ever bumped up against. That was his drawback. However, compared with the work of other newspapermen then employed in Nevada, his stuff was in a class by itself and was commercially very valuable."

AT THIS POINT Rice stated that he was not officially on the staff of the *News*, yet he was blamed for some of its early denunciations against Senator George S. Nixon. Within a week after assuming the editorship, Teague made a front-page attack on Senator Nixon headed: "Goldfield in the Grasp of Wall Street Sharks." Then on May 25, 1907 another editorial declared: "Nixon a Senator with a Blackmailing Mind." These charges certainly were instrumental in causing much of Rice's later troubles.

In June 1907 Teague became editor of the *Nevada State Journal*, thus severing his connection with the *Nevada Mining News* and allowing Rice to succeed as editor.

The Sullivan-Rice partnership finally ended in August 1907. The *Tonopah Miner* reported that Sullivan had assaulted one of the office staff as well as taken the company's books away from the premises. As Rice was then quite ill and unable to prosecute the matter, his wife had initiated a $75,000 suit against Sullivan for damages. The outcome of the case is unknown, but a year later Sullivan was reported to be leaving the country. He was once again without funds and was last seen heading for Mexico to prospect for gold using a $1,000 grubstake obtained from a Los Angeles group.

Previously, the renowned comedian Nat C. Goodwin, while on tour with his co-star Edna Goodrich, had given a performance in Goldfield. Goodwin took such a liking to Nevada and its mines that he decided to enter the mining business as a promoter.

He and a young mining man named Dan Edwards formed a partnership at Reno and came to Rice for advice on investments. Rice gave Edwards a tip on Goldfield Consolidated stock, telling him to sell short. When the stock fell from $7.50 in August to $6.50 in October, Edwards asked Rice to join Goodwin and himself. They raised $10,000 to begin operations and within a week Nat C. Goodwin & Company was incorporated. Goodwin was installed as president, another young mining man named Warren A. Miller was chosen vice-president, and Dan Edwards became secretary. The new company then engaged Rice on a salary to show them how to operate.

It was at this time that the infant mining camp of Rawhide began to make news. Rice visited there and was instantly taken by the possibilities he saw. "I resolved to 'press-agent' the camp," Rice modestly recalled. As editor of the *Nevada Mining News* he helped to bring millions of dollars of investment capital into Rawhide and to promote it into one of the great Nevada boom camps.

Rice, Goodwin and fight promoter Tex Rickard even persuaded Elinor Glyn, author of the best-selling novel *Three Weeks*, to visit the camp, then made certain that she was greeted by a realistically staged gunfight. One of Rice's greatest feats was the acquisition of the Rawhide Coalition Mining Company by the Goodwin company, E. W. King and Eugene Grutt.

Nat C. Goodwin & Company was involved in a number of mining stock ventures, with George Graham Rice firmly orchestrating every move. Finally Goodwin quit the mining game and withdrew from the company in November 1909, bringing about the dissolution of the firm.

One of Rice's best friends during the Rawhide days was Eugene Grutt, one of the four highly respected Grutt brothers who did so much in the legitimate development of Rawhide. When he left Reno for New York in October 1908, Rice made Grutt a present of a barrel filled with fine crystal.

ON JANUARY 18, 1909 Rice became manager of publicity and promotional enterprises for B. H. Scheftels & Company in New York. Within five months its *Scheftels Market Letter* was supplemented by the *Mining Financial News*, the same paper which Rice had worked for when it was published in Reno as the *Nevada Mining News*.

Rice managed to make a great deal of money for his employer until September 29, 1910 when the B. H. Scheftels & Co. offices were raided by U.S. postal authorities. The constant use of mail solicitations to Rice's "sucker" list finally caught up with him.

Ever faithful to the man who had helped to put Goldfield on the map, the October 8, 1910 *Goldfield News* declared:

> Notwithstanding the fact that the government, at the recommendation of Senator Nixon, has been working for three years in an effort to secure evidence tending to show that Rice's enterprises were not legitimate, they have not been successful; every claim made by Rice in regard to any of the properties in which he is interested has been borne out by investigation made by men biased against him. . . . The real person to whom the assaults have been leveled is George Graham Rice . . . and it is known that at least one man high in the government at Washington has been trailing him with postal inspectors for the past three years, using the machinery of the government for his own personal spite and vindictiveness.

Three months earlier, on July 28, the same newspaper had carried a story announcing Thomas G. Lockhart's sale of his controlling interest in the

Jumbo Mining Company to Charles Herzig, Rice's brother and supposedly a noted geologist in his own right.

The Jumbo company owned the Gold Wedge, Paloverde, Vinegarroon and Dick Bland fractions, as well as a group of claims in the adjacent Diamondfield area. The Paloverde and Vinegarroon properties were right in the middle of the rich property of the Goldfield Consolidated Mine Company controlled by George Wingfield. This was probably one of the main reasons for Rice's trial, as Wingfield wanted control of these claims. No doubt George Graham Rice engineered the deal through his brother as the *Goldfield News* of August 6, 1910 noted that Jumbo Extension stock was being heavily boosted in the East. It was also during this period that Rice, in combination with others, carried on extensive mining stock operations in the Porcupine and Dome Mining Districts in Canada.

The Goldfield News during July and August 1910 reported the coming of Charles Herzig and S. C. Constant, alleged to be one of the foremost mining engineers in the world. After visiting Goldfield, Rawhide and Ely, the men were filled with praise for the Rawhide Coalition, Jumbo Extension and Ely Central mines. Yet when it became known that Charles Herzig was Rice's brother, Rice-controlled mining stocks suffered a rapid decline.

But in Goldfield some people still felt that Rice was being unjustly accused. On August 13 the *Goldfield News* editorialized:

> During the past two weeks, there has been a material crystallization of sentiment regarding the return of G. G. Rice and associates to the Goldfield district for his active mining operations and for extensive boosting of the Goldfield district and leases of merit. . . . The general expression is that Goldfield has been too long without people of the ability and the hustle to place Goldfield in its proper light. . . . We ought to have several sets of boosters like the Rice crowd operating in the Goldfield district . . .

Federal authorities were not the only ones unhappy with Rice's activities. The *Goldfield Daily Tribune* reported on February 2, 1911 that he had recently been divorced in Reno. "I am only one of the innumerable dupes of George Graham Rice," his wife complained. Yet the newspaper reported that "she sat in court this morning wrapped in a couple of thousand dollars' worth of furs, and told how she had been rendered a pauper by her husband."

BETWEEN HIS 1910 indictment for using the mails to defraud and the beginning of his trial in the federal district court in New York City, Rice wrote a series of articles which appeared in *Adventure Magazine* entitled "My Adventures With Your Money."

Not all Goldfield newspapers considered Rice to be an asset to the region. An irate *Goldfield Tribune* commented on April 29, 1911:

> The editor of the *Adventure Magazine* . . . says that the enemies of George Graham Rice call him the most malign influence of the world of Get-Rich-Quick Finance. He further says that both friends and enemies agree that George Graham Rice is a master necromancer in the art of inducing small investors to take long chances. George Graham Rice, whose real name is Jacob Simon Herzig, is now under indictment for fraud, but that does not prevent him from writing for the *Adventure Magazine* his life story . . . his impudent re-hash of swindling schemes in which he played the stellar role.

Soon even more biting accusations were appearing in print. On July 1, 1911 the *Goldfield News & Weekly Tribune* quoted from the *Financial World of New York:*

> It is apparent to a close observer that the underlying object of George Graham Rice, in his narrative now appearing in the *Adventure Magazine*, "My Adventures With Your Money", is in preparation of a defense for use when he is brought to the bar of justice to answer for his questionable operations in the field of finance. . . . Throughout his whole career in Goldfield and in New York, Rice was surrounded by blackmailers who were pensioners on his payroll and whom he could not drop for the very good reason that they knew too much about him and might squeal . . .
>
> Rice says Senator Nixon of Nevada was the man who handled the Southern Pacific money in that state to buy crooked legislature, and George Wingfield, he declares, was a crooked gambler, whereas Rice was the embodiment of honesty. . . . On their shoulders he puts the whole responsibility for the downfall of his Sullivan Trust Co.
>
> Compared with Rice and his figurehead partner, "Larry" Sullivan, these men, even if all he says be true, are angels. They may have been rough men and gamblers originally, not an uncommon occurrence in a new mining camp, and it is only to their credit that they have arisen above their earlier environment and become influential citizens of the state. But Rice never rose above the level of a financial confidence man, whose whole record of failure is sufficient proof of this statement.

INTRODUCTION

IN AUGUST 1911, a Grand Jury in New York City handed down an indictment against Rice, Scheftels, Rice's brother Charles Herzig and other employees of the firm. The trial began on October 23, 1911 before Judge Ray in the United States District Court on the charge of using the mails to defraud. More than four months later, on March 7, 1912, Rice and Scheftels finally ended the lengthy trial by changing their pleas to guilty. Rice was sentenced to a one year term in the federal penitentiary at Blackwell's Island but Scheftels was only given a suspended sentence.

Some newspapers commented that after five months of trial Rice was worn out and completely without funds. The *Goldfield Daily Tribune* of March 20, 1912 reported that he had filed a petition of bankruptcy claiming debts of $487,406.00 and assets of only $150 in available cash. In an attempt to gain additional income from his writings, Rice's articles were published in book form in 1913 by the Gorham Press in Boston.

The text is engaging and convincing—to the extent that an uninformed reader could well believe all the statements contained therein to be factual. But Rice's firm conviction that most people are born suckers and that they will always take a chance to reap large rewards with small investments is evident right in the dedication.

Rice may have sold millions of dollars in mining stock to "American Damphool Speculators," but he couldn't sell many books about how he accomplished it. Perhaps the previous magazine accounts had saturated the market, or perhaps events just moved so fast in western mining circles that no one wanted to read of such stale events. Whatever the case, few original copies of the book were sold and they are quite scarce today.

Rice's 1912 conviction and sentencing to the federal penitentiary at Blackwell's Island was certainly not the end of his somewhat unscrupulous career. In fact, he was just hitting his stride.

The *Engineering and Mining Journal* of March 1, 1919 noted that:

> George Graham Rice, a poet of surpassing imagination when it comes to promoting oil and other stocks that are warranted to make purchasers wealthy even faster than a grand jury can return indictments, was adjudged insolvent by Judge Martin T. Manton of the Federal District Court, February 3, in the face of his claim that he had "Rice oil stocks" of great value.... Two indictments of grand larceny against Rice were found by the Grand Jury last July, two days after he was arrested by Federal agents on a charge of using the mails to defraud.

Somehow Rice managed to beat these charges as he had so often done before. However, by early 1920 he was again in serious trouble. The January 31, 1920 *Engineering and Mining Journal* stated that Rice had been convicted in New York of grand larceny, the verdict being reached by the Grand Jury after long deliberation and a sensational trial.

Facts were introduced showing that his company had done business amounting to more than $35 million in 1917 and 1918. Rice testified that he never paid any personal attention to orders of less than $50,000. Evidently, he had far outgrown the 10 to 20-cent mining stocks which he had handled earlier in Goldfield.

Apparently Rice never did serve any time in prison following his 1920 conviction. The *Engineering and Mining Journal* of July 30, 1921 carried an editorial headed "The Awaking":

> The public seems to be awakening to the fact that George Graham Rice is not in jail after all. . . . On January 19, 1920 he was sentenced to three years in Sing Sing prison . . . but he speedily secured his release upon a certificate of reasonable doubt. . . . There has been no secrecy about Rice's conviction nor about his previous career . . . yet the facility with which he has been able to secure acquiescence to his scheming in Reno, San Francisco and more recently Salt Lake is simply amazing. Recently the state of Utah granted the Bingham-Galena Mining Company a stock selling permit over many protests, made because of Rice's connections. The reason being that a majority of the members of the stock exchange favored the granting . . . presumably because Rice is known to be a good manipulator . . . and the stocks he is connected with usually enjoy, for a time, a sensational rise. . . . You can't keep a good swindler down when there is a public that needs trimming.

WHILE RICE'S TROUBLES with the federal authorities in New York were ongoing, he was master-minding another venture involving the Broken Hills Silver Corporation. Situated at the town of Broken Hills about 18 miles east of Rawhide, this property actually did produce $70,000 between 1913 and 1921. Francis Lincoln's *Mining Districts and Mineral Resources of Nevada* stated that Rice's Fidelity Finance and Funding Company of Reno began to fund the corporation until financial difficulties forced them to reorganize.

This was apparently unsuccessful, as the September 17, 1921 *Engineering and Mining Journal* states that the furniture and office equipment of the Fidelity Finance and Funding Company had been attached. Deciding that discretion was the better part of valor, Rice suspended his operations in Nevada for the time being.

The *Engineering and Mining Journal* of December 10, 1921 reports that G. G. Rice was then at Rocky Bar, Elmore County, Idaho, where he was trying to interest New York capitalists in obtaining an option on some mining property. On March 29, 1922 the *Journal* explained that the Idaho Gold Corporation had taken over some old prospects at Rocky Bar. Rice was able to get Samuel Newhouse, a respected mining man from Utah, to serve as president while Jess Hawley, a Boise attorney, was the resident agent. Following an investigation by Idaho State Mine Inspector Stewart Campbell, where he found the value of the ore to average $35 per ton instead of the claimed $275 per ton, Hawley and Newhouse resigned.

During this period Rice and six others were indicted by a federal grand jury on charges of using the mail to defraud in connection with the Fortuna Consolidated Mining Company, a mysterious firm which no one seemed to have any knowledge of.

Rice was reported to be extremely interested in the Weepah boom in south-central Nevada, the last true gold rush to occur in the United States. The *Nevada State Journal* on April 4, 1927 claimed that he had furnished the money to buy the Belcher Extension mine there. The account further stated that Rice was sending his brother and Walter Harvey Reed from New York to Nevada to investigate the Weepah strike, but it is probable that the boom had begun to evaporate before the two men arrived.

Then on April 16, 1927 the *Engineering and Mining Journal* reported another government offensive to stop the sale of Columbia Emerald Development Corporation stock. In this case Rice and his associates had secured an option for 500,000 shares at 75 cents a share, then the stock was boosted up to $17 a share. Yet according to the *Journal* the assets of the corporation—consisting only of a spoonful of stones of inferior quality, two emeralds and $60,000 in cash—were fictitiously valued at $250,000. Somehow Rice was never indicted in this matter, but another case against him was building.

The *Engineering and Mining Journal* on December 17, 1927 related how Rice worked out a deal for his Idaho Copper Corporation to merge with the legitimate Idaho Copper Company. In this merger Rice was able to choose four of the seven directors, which Idaho Copper Company officials submitted to in order to get badly needed funds for development. Eventually a suit was filed against Rice by the Idaho Copper Company and a receiver was appointed. Rice was charged with misuse of the mails to defraud the public, and the government had carefully prepared a weighty case against him.

RICE HAD PREVIOUSLY employed a superbly competent battery of attorneys including former U.S. Attorney Henry Wise, who had prosecuted him in the 1911 trial; Louis Marshall, America's foremost constitutional lawyer; former New York Governor Charles S. Whitman; and former judge Nash Rockwood. Yet for this trial Rice defended himself, relying solely on a young attorney named David Siegel for some advice.

After a four-week trial, Rice and two other defendants were convicted in December 1928, and this time he was indeed bound for prison. He managed to stay out of jail until November 1929 when his conviction for stock fraud was upheld by the Circuit Court of Appeals.

On November 9 the *New York Times* reported that he was being sought by authorities, and the following day they declared that a nation-wide hunt for the missing man had been launched. In reality Rice was living quietly in New York City pending the outcome of his appeals. He was taken into custody and a week later entered the federal penitentiary at Atlanta to serve a four year sentence.

Rice must have remarried after his 1911 divorce for in February 1930, while he was in prison, he was sued for separation by his wife. Apparently this was a very low key case as no further mention is made of the affair.

Even though he was behind bars, federal authorities were still out to get Rice. In October 1931 he left the Atlanta penitentiary long enough to stand trial in New York on a charge of income tax evasion. Once again he acted as his own attorney, and this time he won an acquittal. He even managed to delay his return to prison for a few weeks by obtaining an order demanding that the government return papers seized in the previous case.

Finally Rice completed his sentence and was released from Atlanta early in 1933. By July he had returned to Reno for a few days, ostensibly to investigate some promising mining properties. He even visited Virginia City on July 3, but apparently found nothing to his liking. On July 21 the *Goldfield News and Weekly Tribune* noted that Rice had decided to operate the Halifax mine in California as well as the Buckskin National property in northern Humboldt County, Nevada.

Beginning in January 1934 Rice began issuing a publication entitled *The Financial Watchtower*. Printed in New York City and mailed to subscribers for $10 per year, the newsletter featured information on prevalent gold stocks such as Homestake, Juneau Alaska, and Goldfield Consolidated. The *New York Sun* stated: "George Graham Rice is Back at the Old Stand. Atlanta Alumnus Resumes Operations as Tipster. . . . In some manner Mr. Rice, *nee* Jacob Simon Herzig, obtained newspaper publication of his plans without references to his past experience."

The *Goldfield News and Weekly Tribune* in February 1934 quoted an interview with Rice in New York:

> "You know" he remarked, "I think that big operators on the street who had me railroaded to the U.S. Prison at Atlanta will be more tolerant now.... They got after me because I bought Idaho Cooper at 20 cents and at one time it got up to $6.25, when the government stepped in. Then it went down again to 56 cents. Well the Allegheny of Morgan's was at $72 and it went down to 37½ cents now, but they haven't pinched him yet. I was absolutely innocent in that Idaho Copper thing, anyway."
>
> "That Al Capone is a fine fellow. Often he said to me, 'George, I would not mind being down here if only those big bankers did not get away with it.' We were there together, you know, Al and I."
>
> "'George,' Al used to say, 'I took 5,000 ex-convicts, gave them good jobs selling beer—$150 a week—and what do I get for it? Now and then one of the tough ones would shoot one of the others, maybe arguing over routes, and that gave me a bad name...' I'd take Al's word quicker than anybody's on the stock exchange..."
>
> Rising from his chair, the hawk-nosed Mr. Rice swayed above white spats, as contagiously optimistic as in the days when he ran Maxim & Gay's horse racing tipping service—which netted him $3,000,000—back in 1903, or the Rawhide Coalition Mining Company in 1908, or Appalachian Oil in 1920, or for that matter Idaho Copper.

Unfortunately, Rice's *Financial Watchtower* only served to create additional problems. One of the conditions of his sentence to Atlanta in 1928 was that he be placed on five year's probation upon his release, with an additional five year's sentence to be served if he violated the terms of probation. In November 1936 federal authorities finally decided that the newsletter which he had been publishing for the previous 35 months violated his probation. Calling it a "puff sheet" and charging fraud in the promotion of International Silver and Gold Corporation stock, a warrant was issued for his arrest.

Apparently Rice knew he could not win this one, for then he dropped from sight. Federal agents were unable to locate him, no further stock promotions were ever laid to his credit, and his name ceased to appear in any major New York newspapers. Only a brief mention in the August 1939 *Mining & Metallurgy* reads:

> In the March "Drift" we casually inquired whether or not anyone knew what had become of the redoubtable George Graham Rice, one of the most famous of mining promoters that flourished in the good old days (as Wall Street brokers recall), or the bad old days (as the S.E.C. remembers). It was not long before we had a note from a Milwaukee reader, J. V. Stayoke, saying that George is now the owner of one of the finest cabarets in Milwaukee, working hard and happy as a lark.

Back in 1911, when he was broke and awaiting trial, Rice made an accurate assessment of those flamboyant days of mining booms and busts and the attendant stock manipulations which accompanied them. "Who profited?," he asked. "The answer is: If anybody, the aggregate. The world has been the gainer. It is richer for the gold, the silver, the copper, and other indestructible metals that have been brought to the surface, as a result of this endeavor, and added to the wealth of the nation. But for the gambling instinct and the promoter who caters to it, the treasure-stores of Nature might remain undisturbed and fallow and the world's development forces lie limp and impotent."

George Graham Rice not only added a colorful chapter to Nevada's early 20th century mining history, but certainly hastened the inevitable day when federal regulation of financial investments assured some measure of honesty.

It always pays to deal with a reliable firm.

WELL established firm can stand any misfortune more calmly than that which tarnishes its good name or intimates by direct charge or innuendo that its business methods do not conform to the code of recognized, upright commercial bodies. We have always striven to conduct our affairs along the lines of the Golden Rule, and feel that therein is largely due our success.

As we remarked previously, it is a great struggle for preferment in these busy days of competition in commercialism. We have endeavored to build up and maintain our business by honest, earnest effort and courteous treatment. We have aimed to keep in front.

AS in all walks of life, business or social, there is the black sheep. So it is with mining. The unscrupulous sharper is there. He is practising his shell game. But systematic action is to be taken by the reputable firms to oust these conscienceless thieves. The U. S. government also has its eye out, and the future pathway of this band of robbers is to be made a thorny one.

Goldfield gaming scene, about 1907

Foreword

You are a member of a race of gamblers. The instinct to speculate dominates you. You feel that you simply must take a chance. You can't win, yet you are going to speculate and to continue to speculate—and to lose. Lotteries, faro, roulette, and horse race betting being illegal, you play the stock game. In the stock game the cards (quotations or market fluctuations) are shuffled and riffled and stacked behind your back, after the dealer (the manipulator) knows on what side you have placed your bet, and you haven't got a chance. When you and your brother gamblers are long of stocks in thinly margined accounts with brokers, the market is manipulated down, and when you are short of them, the prices are manipulated up.

You are on guard against the get-rich-quick man, and you flatter yourself that you can detect his wiles at a glance. You can—one kind of get-rich-quick operator. But not the dangerous kind. Modern get-rich-quick finance is insidious and unfrenzied. It is practised by the highest, and you are probably one of its easy victims.

One class of get-rich-quick operator uses crude methods, has little standing in the community, operates with comparatively small capital, and caters to those who do not think and have only small resources. He is not particularly dangerous.

The other uses scientific methods—so scientific, indeed, that only men on the inside readily recognize them; occupies a pedestal in the community; is generally a man of excellent financial standing, a member of a stock exchange; employs large capital; appeals to thinkers or those who flatter themselves that they know the difference between a gold bar and a gold brick, and seeks to separate from their money all classes and conditions of men and women with accumulations large or small.

The United States government during the past few years, at the behest of the big fellow, who seeks a monopoly of the game, has been raiding the little fellow—the crude operator whose power to injure is as nothing compared to the ravages that have been wrought by the activities of his really formidable prototype.

I have a message to communicate to every investor and speculator, a story to tell of my experience through the great Goldfield, Bullfrog, Manhattan and Greenwater mining booms in Nevada of 1905-1908, in which the public lost upwards of $200,000,000, and of a series of great mining stock promotions in Wall Street and other American financial centers, in which the public sank $350,000,000 in 1910. The narration of the facts demonstrates that the government's get-rich-quick crusade has made it less easy for some of the small offenders to thrive, but that the transcendentally greater culprits are at this very moment plucking the public to a fare-you-well, and that the government has not lifted a finger against them.

No man, except a common thief, ever started out to promote a mining company or any other company that he was convinced at the outset had no merit; and the work of common thieves is quickly recognized and the offenders are easily apprehended.

The more dangerous malefactors are the men in high places who take a good property, overcapitalize it, appraise its value at many times what it is worth, use artful publicity and market methods to beguile the thinking public into believing the stock is worth par or more, and foist it on investors at a figure which robs them of great sums of money. There are more than a million victims of this practice in the United States.

After years of experience behind the scenes, the conclusion is forced upon me that the instinct to speculate is so strong in American men and women that they choose to take a chance regardless of the fact that at the outset they already half realize they eventually must lose.

Myself, in boyhood, a victim of the instinct to speculate, I, years afterward, at the age of 30, learned to cater to the insatiable desire in others. I spent fortunes for advertising and wrote my own advertisements. I constructed on big lines powerful dollar making machinery that succeeded in getting the money for my enterprises, and I was generally my own manager. Ten years of hard work in a field in which I labored day and night has disclosed to me that the instinct to gamble is all-conquering among women as well as men—the rich and the poor, the young and the old, the wise and the foolish, the successful and the unsuccessful.

Worse, if you have lost some of your hard-earned money in speculation, your case is undoubtedly incurable, because you have a fresh incentive, namely, to get even. Experience, therefore, will teach you nothing. The professional gambler's aphorism, "You can't kill a sucker," had its genesis in a recognition of this fact, and now stock promoters and manipulators of the multi-millionaire class subscribe to its truth and on it predicate their operations.

Nearly everybody speculates or gambles; few win. Where does the money go that is lost? Who gets it?

Are you aware that in catering to your instinct to invest, methods to get you to part with your money are so artfully and deftly applied by the highest that they deceive you completely? Could you imagine it to be a fact that in nearly all cases when you find you are ready to embark on a given speculation, ways and means that are almost scientific in their insidiousness have been used upon you?

What are these impalpable yet cunningly devised tricks that are calculated to fool the wisest and which landed you? I narrate them herein.

What are your chances of winning in any speculation where you play another man's game? Have you any chance at all?

In playing the horse races in years past you had only one chance if you persisted—you could lose.

In margin trading on the New York Stock Exchange, New York Curb, Boston Stock Exchange, Boston Curb, Chicago Board of Trade, Chicago Stock Exchange, New York Cotton Exchange and kindred institutions, experience among stock brokers proves that if you stick to the game you have only one chance—you can lose.

In railroad, industrial and mining stock speculation, where you buy the shares outright and hold them for stock market profits, you have two chances; if you are of the average and your operations are for a period continuous—you can break even if you are very lucky, or lose if you are not; and in justice to myself I must be allowed to explain that I had a much better opinion of the public's chances ten years ago than I have now, and that experience on the inside has taught me this.

The moral to the investor and speculator is "never again!" And yet you *will* speculate again. Experience teaches that so long as the chance of speculative gain exists in any enterprise, so long will the American public continue in its efforts to appease its speculative appetite.

—GEORGE GRAHAM RICE

Table of Distances.
Hawthorne to Tonopah
100 Miles.
Tonopah to Goldfield
28 Miles.
Tonopah to Gold Reed
70 Miles.
Tonopah to Silver Peak
40 Miles.
Tonopah to Belmont
50 Miles.
Tonopah to Reveille
70 Miles.
Goldfield to Lida
35 Miles.
Goldfield to Gold Crater
35 Miles.
Goldfield to Bull Frog
70 Miles.
Goldfield to Rhyolite
73 Miles.
Goldfield to Sylvania
70 Miles.
Goldfield to Borax Works
130 Miles.
Goldfield to Ash Meadows
112¼ Miles.
COPYRIGHT APPLIED FOR BY
BOOKER PHILBRICK AND FENNER, Jan. 1905.

PROSPECTORS' MAP

OF

Tonopah, Goldfield, Kawich
Bull Frog
and Principal Mining Districts.

Showing Towns, Roads, Trails, and
Watering Places in Southern Nevada
and Death Valley Region, California

BY BOOKER, PHILBRICK & FENNER,
Consulting Mining Engineers,
U. S. Deputy Mineral Surveyors,
GOLDFIELD, NEVADA.

1. The Rise and Fall of Maxim & Gay

THE PLACE WAS NEW YORK. The time, March 1901. My age was thirty. My cash capital, tightly placed in my pocket, was $7.30, and I had no other external resources. I was a rover and out of a job.

Since August of the year before I had been loafing. My last position, seven months before, was that of a reporter for the New Orleans *Times-Democrat*. My last newspaper assignment was the great Galveston cyclonic hurricane in which 15,000 lives were lost and $100,000,000 in property was destroyed. I covered that catastrophe for the New York *Herald* and other journals as well as for the New Orleans newspaper. It was a beat and I netted a big sum for a few days' hard work, but the money had all been spent for subsistence.

At the corner of 40th Street and Broadway I met an old time racetrack friend, Dave Campbell. His face wore a hardy, healthful hue, but he bore unmistakable evidence of being down on his luck.

"Buy me a drink," he said.

"I've got thirty cents in change and I must have a cigar," I answered, "and you know I like good ones."

"Well, I'll take a beer," he said, "and you can buy yourself a perfecto."

No sooner said than done. The cigar and the drink were forthcoming. We sat down. It was a café with the regulation news ticker near the lunch counter.

"Do you still bet on the horses?" asked Campbell.

"No, I haven't had a bet down in more than a year," I answered.

"Well, here's a letter I just received from Frank Mead at New Orleans, and it ought to make you some money," he said.

"There's a 'pig' down here named 'Silver Coin,' " the letter said, "that has been raced for work recently. I think he's fit and ready and that within the next few days they will place him in a race that he can win, and he will bring home the coonskins at odds of 10 to 1."

I had seen letters like that before, but my interest was aroused. I picked up a copy of the New York *Morning Telegraph* from the table. Turning the pages, I noticed a number of tipsters' advertisements, all claiming they were continually giving the public winners on the races.

THE BIRTH OF AN IDEA TO COIN MONEY

"Do these people make money?" I asked Campbell. "Yes, they must," he answered, "because the ads have been running every day for months and months."

"Well, if poorly written ads like these can make money, what would well written ads accomplish, and particularly from an information bureau which might give real information?" I queried. A moment later the ticker began its click, click, click.

"Here come the entries," said Campbell.

He went to the tape and ejaculated, "By Jiminy! Here's 'Silver Coin' entered for tomorrow."

The coincidence stirred me.

"I've got an idea for an advertisement," I said. "Get me a sheet of paper."

It was supplied. I wrote:

> Bet Your Last Dollar On
> "SILVER COIN"
> TODAY
> AT NEW ORLEANS
> He Will Win At 10 to 1

And then I faltered. "I must have a name for the signature," I said.

I picked up the newspaper again and turned to the page containing the entries for that day at the New Orleans races. A sire's name was given as St. Maxim.

"Maxim!" I said. "That's a good name. I'll use it. Now for one that will make euphony."

"Gay!" said Campbell. "How's that? It's sporty."

Thereupon I created the trademark of Maxim & Gay.

In a postscript to this advertisement I stated that the usual terms for this information were $5 per day and $25 per week, and that the day after next Maxim & Gay would have another selection, which would not be given away free.

"Maxim & Gay" were without an address. Half a block away on Broadway, at a real estate office, we were informed that upstairs they had some rooms to let. I engaged one of these for $15 a month—no pay for a week. Two tin signs were ordered painted, bearing the inscription, "Maxim & Gay." One was placed at the entrance of the building and the other on the door upstairs. The sign painter extended credit.

Before bidding me adieu, Campbell all of a sudden, "By golly! I can't understand that scheme. How can you make any money giving out that 'Silver Coin' tip for nothing?"

"Watch and see!" I said.

Around to the *Morning Telegraph* office, then on 42nd Street, I went. "Insert this ad and give me $7 worth of space," I said, as I shelled out my last cent.

When the advertisement appeared the next morning, its aspect was disappointing. The space occupied was only fifty six agate lines, or four inches, single column measure. It looked puny. Would people notice it?

That afternoon Campbell and I took possession of the new office of Maxim & Gay. Luckily, a former tenant had left a desk and a chair behind, in lieu of a settlement for rent. In walked a tall Texan.

"Hey there!" he cried. "Here $5. It's yours. Keep it. Answer my question, and no matter what way you answer it, it don't make any difference. The $5 is yours."

I looked up in amazement.

"Give me the source of your information on 'Silver Coin,' " he said. "I bet big money. If your dope is on the level, I'll bet a 'gob.' If it ain't, your confession will be cheap at $5, which will be all the money I'll lose."

I showed him the letter from Frank Mead.

"That's good enough for me," he said, turning on his heel. "Silver Coin" won easily at 10 to 1.

The betting was so heavy in the New York pool rooms that, at post time, when 10 to 1 was readily obtainable at the race track, 6 to 1 was the best price that could be obtained in New York. It is history that the New York City pool rooms at that time controlled by "Jimmy" Mahoney were literally burned up with winning wagers. Pool room habitués argued it thus: "If the tip is not a good thing, what object in the world would these people have for publishing the ad? If the horse loses, the cost of the advertisement is certainly lost. The only way they can win is for the horse to win." It was good logic— as far as it went.

THE HIGHER MATHEMATICS OF THE OPERATION

But it was really sophistry. If the horse lost, the inserter of the Maxim & Gay advertisement would be out exactly $7. If the $7 was used to bet on the horse, the most that Maxim & Gay could win would be $70. I was taking the same losing risk as the bettor, with a greater chance for gain. By investing $7 in the advertisement, it was possible for me to win much more money from the public by obtaining their patronage for the projected tipping bureau.

I recall that the experimental features of the advertisement appealed to me strongly and struck me as being a splendid test of the possibilities of the business. If the horse won and there were few responses to the advertisement it would be convincing on the point that there was no money in the tipster branch of the horse racing game. I argued that if the racing public would not believe that an Information Bureau was what it cracked itself up to be, in the face of a positive demonstration, how could it be expected to believe the lurid claims of the fakers whose advertisements crowded the sporting papers daily and in which they claimed *after* the races were run that they named in advance the winners at all sorts of big odds?

The next morning about ten o'clock, Campbell called at my home and said that he had received another good thing by telegraph from Mead and that the name of the horse was "Annie Lauretta," with probable odds of 40 to 1.

"Jiminy!" he exclaimed. "If we only get a few customers today and this one wins, what will happen?"

Leisurely we walked to the office. "If we get ten subscribers today to start with, we'll make a fine beginning," I said.

As we approached the Hotel Marlborough, which is opposite the building on Broadway in which the Maxim & Gay Co. had its modest little office, our attention turned abruptly to a crowd of people who were being lined up by half a dozen policemen.

"What theater has a sale of seats today?" Campbell asked.

"Don't know," I answered.

As we approached the office, we found that the line extended into our own office building. As we ambled up the rickety stairs, we passed the crowd in line, one by one, until we discovered, to our great astonishment, that the line ended at our door.

We turned the key, walked in, locked the door, and stood aghast.

Holding up both hands, I gasped, "In heaven's name, what have we done?" I was appalled.

"Give 'em 'Annie Lauretta,' " cried Campbell.

"But suppose Annie don't win," I expostulated.

"Smokes!" exclaimed Campbell. "Are you going to turn down all those $5 bills?"

"Let's see that telegram," I faltered. I perused it over and over again.

"Mead's judgment on 'Silver Coin' is good enough reason to warrant advising people to put a wager on another one of his choices," Campbell argued. I agreed.

How to convey the information in merchantable form was the next question. A typist in the Hotel Marlborough, across the way, was sent for and asked to strike off the name "Annie Lauretta" 500 or 1,000 times on slips of paper. Envelopes were bought and a typed slip was placed in each. The line increased until it was a block and a half long.

When all was ready, the door was opened. Campbell passed the envelopes out as each man handed me $5. I stuffed the money in the right hand drawer of the desk, and when that became choked, I stuffed it in the left hand drawer. Finally, the money came so thick and fast that I picked up the waste paper basket from the floor, lifted it to the top of the desk and asked the buyers to throw their money into the receptacle. When a man wanted change, I let him help himself.

For two and a half hours, or until within fifteen minutes of the calling of the first race at New Orleans, the crowd thronged in and out of our office. When the last man passed out we counted the money and found the day's proceeds to be $2,755.

"What will we do next?" asked Campbell. "What's my job, and what do I get?"

"How much do you want?" I asked.

"Ten dollars a day," he said.

Thereupon he got possession of the $10 and he admitted it was more money than he had seen in a month.

"What will we do next?" he repeated.

"Let us take a walk," I said. "Lock the office until after the fourth race, when we see what 'Annie Lauretta' does."

We hied ourselves to a nearby resort and stood by the news ticker to see what would happen to Annie. Half an hour passed since the third race report.

"Fourth race—tick—tick—tick," it came. "A—Al—,"

"We've lost!" I cried.

"'A—Al—Alpena' first."

There was grim silence.

"Tick—tick—,"

"Here she is!" yelled Campbell.

"'A-n-n-i-e Lauretta' second—40—20—10" (meaning that the odds were 40 to 1, first; 20 to 1, second; and 10 to 1, third; and that those who had played across the board had won second and third money at great odds).

I boarded a Broadway car, rode down to the Stewart building and rented one of the finest suites of offices in its sacred purlieus. I ordered a leading furniture dealer to furnish it sumptuously. At night I walked over to the *Morning Telegraph* office, laid $250 on the counter, ordered inserted a flaring full page advertisement announcing that Maxim & Gay had given "Annie Lauretta" at 40, 20 and 10, second, and previously "Silver Coin" at 10 to 1, won, and were ready for more business.

A telegram was sent to Frank Mead, instructing him to spend money in every direction with a view to getting the very best information that could be obtained from handicappers, clockers, trainers and every other source he could reach. Mead continued to wire daily the name of one horse, which we promptly and thereafter advertised daily as "the one best bet." Soon "one best bet" became a term to conjure with.

The success of this enterprise was phenomenal. In the course of two years it earned in excess of $1,500,000. There were some weeks when the business netted over $20,000 profits. At the height of its career, in the summer of 1902, at the Saratoga race meeting, when the pool rooms in New York were open, our net profits for the meeting of a little less than three weeks were in excess of $50,000.

We established an office in Saratoga and our average daily sales on race days were 300 envelopes at $5 each. In New York the average was just as large, and, in addition, we had a large clientele in distant cities to whom we sent the information by telegraph. The wire business, in fact, increased to such an extent that it became necessary to call upon the Western Union and Postal Telegraph companies to furnish our office in the Stewart building with direct loops.

I spent the money as fast as I made it. I believed in our own information and made the fatal error of plunging on it. My error, as I afterwards concluded, was in not risking the same amount on every selection. Had I done this, I would not have suffered serious losses. The trouble was that every time a horse on which I wagered won, I was encouraged to bet several times as much on the next one, and by doubling and trebling my bets, I played an unequal game.

The expense of gathering this information within a few weeks increased to upwards of $1,000 a week, and it was not only our boast, but an actuality, that the Bureau did really give more than value received.

Undoubtedly, the evil of the venture was the gambling it incited; but the effort to secure reliable information was honest, and what young man of my age and of my experiences, having indulged in a lark of the "Silver

Coin" variety, could withstand the temptation of seeing the thing through?

Among the leading patrons of the Maxim & Gay Co. were soon numbered important horse owners on the turf, leading bookmakers and many leaders of both sexes in the smart set. Maxim & Gay made it a rule to sell no information of any kind to minors and often excluded young men from the offices for this reason.

HOW "THE ONE BEST BET" WAS COINED

OUR METHODS OF advertising were unique. We used full pages whenever possible, and it was a maxim in the establishment that small type was never intended for commercial uses. We used in our big display advertisements a nomenclature of the turf that had never before been heard except in the vicinity of the stables, and we coined words and phrases to suit almost every occasion. The word "clocker," meaning a man who holds a watch on horses in their exercise gallops, was original with us, and has since come into common use, as has the phrase, "the one best bet," which we also coined.

It was our aim, in using the language of horsemen, to be technical rather than vulgar, the theory being that, if we could convince professional horsemen that we knew what we were talking about, the general public would quickly fall in line.

One morning we were alarmed to see in the *Morning Telegraph,* on the page opposite our own daily effort, the advertisement of a new tipster who called himself "Dan Smith." Dan went Maxim & Gay one better in the use of racetrack terminology. He evidently employed a number of Negro clockers, for the horse lingo which he used in his advertisements smelled of soiled hay and the manure pile. It was awful! But it made a hit with racegoers, and before a week had passed we recognized "Smith" as a dangerous competitor.

We were lead to believe that the use of this horsy language was entirely responsible for Smith's success, for we knew that his tips were not so good as ours. We investigated. His trick was this: In the sheet that he sent out to his customers, he would name for every race at least five horses as having a chance to win. He advised his clients, in varying terms, to bet on every one of them, and if any one of them won, he would print next morning what he had said on the preceding day regarding the winner alone, leading the public to believe that the only horse he had fancied was the actual winner.

I decided to organize another Bureau to knock out Dan Smith. The intention was to go our competitor "a few better" in the use of vulgar horseracing colloquialisms and exaggerated claims, and thus nauseate the betting public and "put the kibosh" on Dan. We created a fictitous advertiser whom we named "Two Spot," and the next morning there appeared at our instigation in the *Morning Telegraph* a large display advertisement, headed substantially as follows:

TWO SPOT
Turf Info. Merchant
Terms, $2 Daily; $10 Weekly

Following the style which Dan Smith had adopted in his racing sheets, "Two Spot" mentioned in his first advertisement, as a sample of his line of dope, four or five horses to win each race, each one in more grandiloquent terms than the other, but these were selected because they, in reality, appeared to be the most likely losers of all the entries.

A woman was sent over to the newly organized office of "Two Spot" to take charge of the salesroom. I was completely taken off my feet the next day when she informed me that the receipts, as a result of the first advertisement, were in excess of $300, and that the public not only did not read between the lines, but had actually fallen for the hoax.

To cap the climax, on the second day one of the "outsiders" which "Two Spot" named derisively as the one best bet "walked in" at 40 to 1! Next day "Two Spot" did a land office business, and within a few days we figured that the "Two Spot" venture would net $1,000 a week if continued. "Two Spot" then went after the game hammer and tongs and endeavored to gage the full credulity of the public.

The distinctive difference between "Two Spot" and Maxim & Gay was this: Maxim & Gay, except in one instance, which is chronicled herein, never pretended to have selected a winner when it had not, while "Two Spot" enjoying the same source of information as Maxim & Gay, worded his daily advices to clients so artfully as to be able to claim the next morning in his advertisements à la Dan Smith, the credit of having said something good about every winner.

The profits of Dan Smith's venture, I was informed, exceeded a quarter of a million dollars the first year, and the profits of "Two Spot," whose career was cut short within a month by a realization on our part that we could not afford to be identified with such an enterprise, was divided among the employees of the "Two Spot" office. "Two Spot" had been brought into being for the purpose of killing opposition and not for profit-making. The scheme failed of its purpose.

To give an idea of the character of some of the raw kind of advertising put out by "Two Spot," and for which the public fell, I recall this excerpt from one of his tipping sheets:

> I am my own clocker. I have slept under horse blankets for thirty years. I understand the lingo of horses. Last night, when I was taking my forty winks in the barn of Commando, I heard him whinny to Butterfly and tell her to keep out of his way today because he was going to tin-can it from start to finish, and if Butterfly tried to beat him, he would savage her. That makes it a cinch for Commando. Bet the works on him to win.

REAL INSIDE TURF INFORMATION

MAXIM & GAY repeated the "Silver Coin" method of advertising only once during the entire career of the company. This happened in the spring of 1902, when John Rogers, trainer for William C. Whitney, sent to the post a mare named "Smoke." Our information was that the mare would win, and our selections for the day named her to win—and she did. Two days later, she was again entered, against an inferior class of horses, and the handicap was entirely in her favor. Notwithstanding this, we inserted an advertisement which appeared in the newspapers on the morning of the race, reading substantially as follows: "Don't bet on 'Smoke' today. She will be favorite, but she will not win. 'Rockstorm' will beat her."

Sure enough, "Smoke" opened up favorite in the betting. The betting commissioners of Mr. Whitney placed large wagers on the horse with the bookmakers. The bulk of the public's money, however, went on "Rockstorm," and before post time thousands of dollars of the wise money followed suit.

"Rockstorm" won the race. "Smoke" led into the stretch, when up went her tail and she "blew up."

Immediately I was cross questioned by messengers from the judges' stand. They asked our reason why we were so positive that "Smoke" would lose. Mr. Whitney, I was informed, was actually suspicious that his mare had been pulled. The reason for the reversal of form, as I explained at the time, was this:

William Dozier, our chief clocker at the racetrack, who had witnessed the preparation which "Smoke" received for the races, was of the opinion that her training had been rushed too fast, and that her first race, instead of putting her on edge, had caused a setback. Her first race, in fact, had soured her. Being a veteran horseman, he was positive that "Smoke" would lose. I afterwards learned that the training of "Smoke" had been left to an understrapper, and that Mr. Rogers himself was not responsible for her condition.

THE PUBLIC ASKS TO BE MYSTIFIED

THE JUDGES WERE apparently satisfied, but the public could not readily understand the truth, and we didn't point it out in our advertisements, because our policy was always to appear as mysterious as possible as to the source of our information.

Mystery played an important rôle in our organization, and it would have been better had we never succeeded in the "Smoke" coup. Up to this time my personal identity had not been revealed at the racetrack, and even the bookmakers did not know who was the guiding spirit of Maxim & Gay. "Jimmy" Rowe, trainer for James R. Keene; Peter Wimmer, trainer for Captain S. S. Brown of Pittsburg, and John Rogers, trainer for William C. Whitney, were at this early period at various times the rumored sponsors for Maxim & Gay. The bookmakers and talent generally conceived the idea that nobody but a very competent trainer in the confidence of horse owners could possibly be responsible for so much exact information regarding the horses. Of course, the track officials who made it their business to know everything knew of my connection with the organization. No sooner, however, did their messengers ask an interview with me than the fact became public property around the racetrack and the mask was off.

The effect for a while was very bad, for our business fell off considerably. "Bismarck" Korn, the well known German bookmaker, put it to me this way on the day of the "Smoke" incident:

"You are the first horse tipster I effer saw dat vore eyeglasses, sported a cane, und vore tailor made cloding. You look like a musicianer—not like a horseman. You're a vonder!"

Gottfried Walbaum, another old-time bookmaker, chimed in: "Dat vas obdaining money under false bredenses. I gafe your gompany dwendy fife dollars a veek for two months alreaty. You gif me my money pack! You are a cheater!"

Riley Grannan, the plunger, said, "Got to hand it to you, kid! Any time you can put one over on the Weisenheimers that have been making a living on racetracks for twenty years you are entitled to medals!"

The attitude of "Bismarck" and of Walbaum was amusing, that of Grannan flattering. But it was poor business, because most of these professional racetrack people ceased for a while to subscribe for the Maxim & Gay service.

For months I had purposely kept myself in the background, fearing a dénouement of this very description. I recalled that in the late 80's, in a town of northern Vermont, when John L. Sullivan was advertised to appear in a sparring exhibition, his manager met him at the train, and, although it didn't rain and the sun didn't shine, an umbrella was raised to cover John L. while walking from the train to a waiting landau. No sooner did Sullivan enter the vehicle than the blinds were drawn. When the carriage reached the hotel, it stopped before a side door. The manager alighted before Sullivan, again quickly raised the umbrella and whisked the heavyweight champion past the crowds and up to his room without exposing him to the view of anybody whatsoever.

Throughout the day Sullivan was screened from public gaze. His face was not seen by a single citizen of the town until he appeared on the stage that night.

I asked the manager why he was so very careful to shield Sullivan from the popular view prior to his appearance before the footlights. I recall that he said:

"If the public thought John L. was just an ordinary human being with black mustaches and a florid Celtic face, they wouldn't go to see him. The public demand that they be mystified, and to have shown people off the stage that Mr. Sullivan is just a plain, ordinary mortal would disillusion them and keep money out of the house."

That piece of showman's wisdom was fresh in mind during the early career of Maxim & Gay; and so long as Maxim & Gay kept racetrack men guessing as to who was directing its destinies, the organization was a howling success. Its good periods were mixed with bad periods after the mystery of sponsorship was cleared up to the satisfaction of the professionals by the inquiry of the racetrack judges into the "Smoke" affair.

A few weeks after the "Smoke" coup, our chief clocker informed us that the entries for a big stake race which would be run on the following Saturday had revealed to him a "soft spot for a sure winner," as he expressed himself, and he said we could advertise the happening in advance with small chance of going wrong. This we proceeded to do.

Money poured in by telegraph from distant cities for the "good thing" on Saturday. Our advertisement on the Thursday previous to the race read like this:

> THE HOG KILLING *of the* YEAR
> Will Come Off at Sheepshead Bay
> On Saturday, at 4 o'Clock.
> *Be Sure to Have a Bet Down.*
> Telegraph Us $5 for the Information

One of our constant patrons resided in Louisville. He was among the first to whom we telegraphed the information on Saturday morning. The race was run and the horse *lost*.

About 4:30 P.M. we received a dispatch from our Louisville customer, reading as follows: "The hog killing came off on schedule time—here in Louisville. I was the hog."

Another message from a pool room habitué reached us, reading: "Good game. Have sent for more money."

We were often in receipt of messages of similar character on occasions when our selections failed to win and our customers lost their money; but these communications were generally in good spirit.

On one occasion we had what we believed to be firsthand information regarding a horse which was being prepared for a big betting coup by Dave Gideon, one of the cleverest horsemen in the country. Following our customary method of using vividly glowing advertisements, with the blackest and heaviest gothic type in the print shop, we announced:

> A GIGANTIC HOG KILLING
> We have Inside Information of a Long Shot
> that Should Win Tomorrow at 10 to 1 and
> Put Half of the Bookmakers out of Business.
> BE SURE to HAVE a BET DOWN on THIS ONE.
> Terms $5.

The *argument* of the advertisement, which appeared beneath these display lines, was couched in the most glowing terms, and made it very plain that our information came from a secret source, and, further, that we had spent legitimately a snug sum of money to secure the information. We also pointed out that the owner was one of the shrewdest betting men on the turf and seldom went astray when he put down a "plunge" bet on one of his own entries.

Next day the race was run. The horse did not finish in the money.

The following day we received many letters, as we always did when one of our heavily advertised good things lost. One of the most unique of these epistles contained a remonstrance from a Philadelphia subscriber. He wrote in this vein:

> Dear Sir:—You have been advertising for some days that you would have a gigantic hog killing today. I was tempted by your advertising bait and fell—and fell heavily with my entire bank roll. My bucolic training should have warned me that

"hog killings" are not customary in the early spring, but I fell anyway.

Permit me to state, having recovered my composure, that Armour or Swift need have no fear of you as a competitor in the pork-sticking line, for far from making a "hog killing," you did not even crack an egg. Pardon me. Thanks. Good-by. Yours truly,
—Subscriber

PRESTIGE RESTORED BY A CLERK'S RUSE

IN THE SUMMER of the second year of Maxim & Gay's great money gathering career, the Information Bureau was out of luck and the patronage of the Bureau fell away to almost nothing. At this period I was seriously ill and confined to my home. A man in my office decided to take advantage of my absence from the scene to improve business a bit on his own hook.

It was the habit of our track salesmen, dressed in khaki, to appear at the office at noon every day and receive a bundle of envelopes containing the tips on the races, and then immediately to proceed to the racetrack, stand outside of the gates and vend them at $5 per envelope.

One day these men, without their knowledge, were supplied with envelopes containing blank sheets of paper instead of the mimeographed list of tips. When a handful of town customers reached the office, they were informed that the selections would be late that day and would be on sale at the track only.

At about half-past one o'clock the 'phone bell rang, and word came from the track messengers that apparently a mistake had been made, as their envelopes contained blanks. They were being compelled to refund money. They asked what to do.

"Wait," they were told. "We will send a messenger immediately with the tips."

The messenger never reached the track. There were no tips issued. On that day "May J." won at odds of 200 to 1.

The next morning, the newspapers contained full page advertisements announcing that Maxim & Gay had tipped "May J." at 200 to 1 as the day's "One Best Bet." It could not have been done without a comeback if any tips had been issued.

A BOASTFUL RACE PLAYER GIVES AID

I WAS NOT present, but I learned as soon as I became convalescent that on the afternoon of the day the advertisement appeared claiming credit for "May J." at 200 to 1, the office was thronged with new customers who enrolled for weekly subscriptions at a rate that put new life into the business. A few of the customers expressed some doubt as to whether Maxim & Gay gave out the 200 to 1 shot or not.

That afternoon there appeared on the scene a race player who, laying $5 down on the desk, said, "Give me your good things. I played 'May J.' yesterday at 200 to 1 and I am rolling in money."

"Where did you buy your information?"

"From your man at the entrance to the track," he answered.

"At what time?" he was asked.

"A quarter to two," he replied.

"Say, young man, there were a lot of people who came in here this morning who said they were not sure we gave out that selection at all. Would you make an affidavit that you bought the information from us?"

"You bet I will!" he said; and thereupon a notary public was called in and the caller swore that he had bought the Maxim & Gay tips at the entrance to the racetrack and that they contained "May J." at 200 to 1.

That affidavit was posted in the office during the remainder of the day. When the clerk who performed this stunt was asked for more information as to how he came to secure such an affidavit, he gave absolute assurance that he did not offer the customer the smallest kind of bride to make it, and that nothing but an innate desire to call himself on top had influenced the man to perjure himself.

But I could not tolerate the misleading advertising that had been done as a result of misplaced energy, and the man responsible for it did not remain with the company.

FORTUNE CHANGES HER MOOD AND SMILES AGAIN

PECULIARLY ENOUGH, the "May J." advertisement was followed by a series of brilliant successes for Maxim & Gay in the selection of winners at big odds, and, within a month our net earnings again reached $20,000 per week. Horse owners, horse trainers and society people who frequented the clubhouse at the racetrack were our steadiest patrons.

The women particularly were most loyal to our bureau. The wife of a young multi-millionaire of international prominence was one of our most ardent followers. She would never think of putting down a bet without first consulting Maxim & Gay's selections. On a notable occasion, this lady arrived at the gate of the Morris Park racetrack with her husband, in their automobile, and took the long stroll to the club house. They were a trifle late for the first race; the horses were already going to the post up the Eclipse chute.

Suddenly the lady discovered she had forgotten to purchase Maxim & Gay's selections. Hastily calling her husband, she gave him a sharp berating for not reminding her to buy the selections. They had a short but earnest interview, which was suddenly terminated by the young man doing a sprint of a quarter of a mile down the asphalt walk from the clubhouse to the main entrance where the tips were sold by the uniformed employees of Maxim & Gay.

Those who witnessed the sprint of the young financier attested to the fact that he never showed as much swiftness of foot in his early college days; but even his unusual speed failed to get him back on time to acquaint his wife with the name of the horse selected by Maxim & Gay for the first race, the race having been run and the Maxim & Gay selection having won. The gentleman thereupon got a curtain lecture from his better half that astonished and amused the society patrons on the clubhouse balcony. Thereafter, he never forgot to get the Maxim & Gay selections. In fact, he made assurance doubly sure by engaging the colored attendant in charge of the fieldglasses to deliver the selections to him daily immediately upon his arrival at the course.

Our popularity with racehorse proprietors was mixed. Among the horse owners with whom we transacted business was Colonel James E. Pepper, the late noted distiller and owner of a big breeding farm and a stable of runners. He was an ardent lover of horses, and maintained that his native Kentucky knowledge of thoroughbreds afforded him an opportunity to pick probable winners of horseraces better than any of "them —— faking tipsters." He had great confidence in his judgment for a while.

THE KENTUCKY COLONEL FALLS IN LINE

AFTER SEPARATING HIMSELF from much cash, while one of his very intimate friends was cleaning up plenty of money on our selections, he finally strolled into our office one morning and sheepishly stated that one of his fool friends had asked him to step in and get our fool selections for him. We explained that it was against our rule to give out our choices before 12:30 P.M., whereat he grew exceedingly wroth. He finally agreed to our conditions, paid his money and was given an order to get the selections at the track entrance from one of our messengers.

Nearly all of our choices won that day. Colonel Pepper came in the following morning and paid for another subscription, this time for a week's service. We were in our stride, the majority of our selections winning from day to day, and Colonel Pepper had cause for exultation. On one of these days we divulged, on our racing sheet, the name of a sleeper that we were confident would win at 10 to 1, a big betting coup having been planned by that Napoleon of the turf, John Madden. The horse won at big odds, and Colonel Pepper made a killing on the information.

For the next day, our clockers had spotted another horse that had been got ready by the light of the moon, and we spread it pretty strong in our advertisements that the horse we would name could just fall down, get up again and then roll home alone. The horse did not fall down; but he won; he rolled home alone by about ten lengths. He belonged to Colonel Pepper. It was anticipated that about 20 to 1 would be laid against this fellow, but on account of our strong tip, he opened at 10 to 1 and was played down to 3 to 1. The bookmakers were badly crimped.

The next day, as soon as the office opened, Colonel Pepper, hotter under the collar than even his name might indicate, stamped into the outer room. Slamming his can down on the big mahogany table, he demanded in stentorian tones: "What in the ——— does this ——— business mean? Here I come and subscribe my good money to your ——— fool tips, and you all are so low down mean as to give my hoss for the good thing yesterday! What does it mean, suh; what does it mean?"

The use of considerable diplomacy was necessary to calm down the irate Colonel, who had no compunctions in winning a big bet on Mr. Madden's "sleeper," but "——— it, suh, it is outrageous to treat *me* so."

The Colonel never got over that incident, and while he won a big bet on his own horse, he always claimed that Maxim & Gay had ruined the betting odds for him and that but for the vigilance of our clockers his winnings would have been twice as large. This was true, and time and again we ruined the price for many another owner who thought he was going to get away with something on the sly.

Bookmakers as a rule are very much self-satisfied about their knowledge of the mathematics of the game. In order to show them that they didn't know all about it, the Maxim & Gay Co. inserted an advertisement one day reading substantially as follows:

YOU PAY US $5
WE REFUND $6
If the Horse We Name as
The One Best Bet
Today Does Not Win, We Will Not
Only Refund Our $5 Fee, Which Is
Paid Us for the Information, but Will
Pay Each Client an EXTRA DOLLAR

The Rise and Fall of Maxim & Gay ☆ 45

> By Way of Forfeit.
> Pay Us $5 Today for Our One Best
> Bet, and if the Horse Does Not Win
> We Will Pay You $6 Tomorrow.
> —Maxim & Gay Co.

Our receipts that day were approximately $5,000. The horse did not win. We refunded $6,000 next day, and netted a considerable sum of money on the operation.

It happened to be a two horse race. Our horse was at odds of 1 to 6 in the betting, that is to say, the bookmakers laid only one dollar against every six bet by the public. The other horse ruled at odds of 5 to 1, meaning that here the bookmakers laid five dollars against the public's one.

The Maxim & Gay Co. sent to the track $1,000 out of the $5,000 paid in by its customers and wagered the $1,000 on the contending horse at odds of 5 to 1, drawing down $4,000 in winnings. From this money it paid its clients the thousand dollar forfeit, netting $4,000 on the operation, after of course returning to them their own $5,000.

Had the 1 to 6 shot won, the clients who had received the winning tip would have been happy, while the Maxim & Gay Co. would not have been compelled to refund any money and would have been ahead $4,000 on the operation, the $1,000 wagered and in that event lost in the betting ring on the other horse being subtracted from the $5,000 paid in by its customers. No matter which horse won our gain was sure to be $4,000 and we had here the ideal of a sure thing.

It was a case of taking candy from a baby; and yet many of the wise bookmakers could not at first figure it out. Nearly all of them subscribed for the information. As for the public, they did not seem to catch on at all.

BETTING THE PUBLIC'S MONEY AT GREAT PROFIT

The Eastern racing season was about to close and it was decided to remove the entire force of clerks to New Orleans for the Winter and there to depart from the usual practice of selling tips only, and to bet the money of the American public on the horses at the racetrack in whatever sums they wished to send. The company employed Sol Lichtenstein, then the most noted bookmaker on the American turf, to bet the money, and made him part of the organization, giving him an interest in the profits.

The Maxim & Gay Co. at this time had made close to $1,000,000, and recklessly and improvidently I had let it slip through my fingers. It was easy come and easy go. As I review that period in my career, I recall that the whole enterprise appeared to me in the light of an experiment—just trying out an idea, and having a lot of fun doing it. Because of its dazzling success I became so confident of my ability to make money at any time that I didn't take serious heed whether I accumulated or not. Besides, I had never loved money for money's sake. All the pleasure was in the accomplishing.

The races at New Orleans were advertised to start on Thanksgiving Day. On the 15th of October I ordered $20,000 worth of display advertising to run in thirty leading newspapers in the United States four days a week, until Thanksgiving. Credit was extended for the bill by one of the oldest advertising agencies in America.

The advertisements told the public to send their money to Maxim & Gay, Canal Street, New Orleans. On my arrival there, two days before Thanksgiving, I called at the post office, and asked if there was any mail for Maxim & Gay. The post office clerk appeared to be startled. He gazed at me as if he were watching a burglar in the act. His demeanor was almost uncanny. He didn't talk. He didn't even move. He just looked. Finally I asked, "What is the matter?"

"Wait a minute," he muttered.

He left the window. He did not return. Instead, what appeared to me to be a United States deputy marshal ambled up to my side and said, "See here; the Postmaster wants to see you."

I was escorted into a secluded chamber in the post office building, and a few minutes later a post office official, along with three or four assistants, came into the room.

"What's the trouble?" I asked.

"You bring us a recommendation as to who you are and what you are and all about yourself before we will answer any of your questions as to how much mail there is here for you," the official said.

I smiled. The advertising, then, was a success.

Having been employed as a newspaper man in New Orleans a few years before, I knew one of the leading lawyers of the city and several bank officials. Within thirty minutes I had lawyer and bank men before the Postmaster, vouching for my identity. Thereupon I was informed that there were 1,650 pieces of registered mail, evidently containing currency, and, in addition, twelve sacks of first class mail matter, which contained many money orders, checks and inquiries. The official said that in the money order department they had notices of nearly 2,000 money orders issued on New Orleans for the Maxim & Gay Co.

I sent a wagon for the mail, and notwithstanding the fact that a force of four men under me opened the letters and stayed with the job for two days, the task was not completed when the first race was called on Thanksgiving Day. On adding up the receipts, we found a little over $220,000.

The meeting continued 100 days, and our total receipts for the whole period were $1,300,000.

Maxim & Gay's system of money making at New Orleans was as follows: we charged each client $10 per week for the information. We charged 5 per cent of the net winnings in addition, and we further contracted to settle with customers only at the closing odds for bets placed, retaining for ourselves the difference between the opening odds and the closing odds. The profit averaged approximately $7,000 a day for 100 days—to us.

As a guarantee of good faith, the Maxim & Gay Co. agreed with its clients that each day it would deposit in the post office and mail to them a letter bearing a post mark prior to the hour of the running of the race, naming the horse their money was to be wagered on; and this was always done. An honest effort, too, was always made to pick a horse that was likely to win, for even a child can see that if we did not intend to bet the money and wanted to pick losers, all we would have had to do was to make book in the betting ring at the racetrack and not spend thousands of dollars in advertising for money to lay against ourselves.

Did we invariably bet the money of our clients on the horse we named?

Yes, always—except once!

$130,000 IS LOST AND WON IN A DAY

THAT INCIDENT is not easily forgotten by several. On this day the entry which we selected was one of Durnell & Hertz's string. The horse was known to be partial to a dry track. The dope said he could not win in heavy going. It was a beautiful sunshiny morning when we selected this horse to win, and at noon the envelopes containing the name of the horse were mailed in the post offices, as usual.

Something happened.

Half an hour before the race was run it began to rain in torrents and the track became a sea of mud. Durnell & Hertz, realizing that they were tempting fate to expect their horse to win under such conditions, appeared in the judges' stand and asked permission to scratch their entry. The judges refused. I asked Sol Lichtenstein, who had the wagering of our client's money in charge, what he proposed to do about betting on the horse under the changed conditions. He exclaimed, "Bet? Do you want to burn up the money?"

"Well, if he wins," I replied, "we will have to pay, because if he wins and you don't bet and we say we changed the selection on account of the rainstorm, they will not believe us and we will have trouble."

"Very well," he said. "You bet my book all the money, and we will, for the first time, book against our own choice. It's fair, because we must pay if we lose, and there is no way out of it. But don't burn up that money." I agreed.

The opening odds against the horse were 2 to 1. Had it been a dry track, he would have opened a hot favorite at 4 to 5 or so. Slowly the odds lengthened to 10 to 1, which was the ruling price at the close. Durnell & Hertz bet on another horse to win. Standing before Sol Lichtenstein's book, I said, "Thirteen thousand on our selection, Sol."

"One hundred and thirty thousand to $13,000," he answered. "Here's your ticket."

Sol and I repaired to the press stand to see the race. Durnell & Hertz's entry got off in the lead. At the quarter he was in front by two lengths. At the half the gap of daylight was five lengths. At the turn into the stretch the horse was leading by nearly a sixteenth of a mile. Then I heard a noise behind me as if a miniature dynamite bomb had exploded. Sol's heavy field glasses had dropped to the floor.

Sol did not wait to see the finish. The horse won in a gallop. At the office of Maxim & Gay accounts were figured and checks signed for the full amount of our obligations, and they were immediately mailed to all subscribers.

At midnight I met Sol in the St. Charles Hotel lobby. He looked worn. "I guess that will hold us!" he moaned.

"Hold us?" I answered. "Nothing better ever happened. It'll make us!"

"You poor nut!" he exclaimed. "Lose $130,000 in a day and it will make you! Stop your noise!"

"Listen!" I rejoined. "At an expense of $3,000 for tolls I have telegraphed a full page add to fifty leading city newspapers, telling the public that we tipped this horse today at 10 to 1 and that we mailed checks to our customers tonight for $130,000. The gain we will reap in prestige and fresh business will repay our loss on the horse."

The next day the Western Union Telegraph Co. found it necessary to assign three cashiers to the work of issuing checks to the Maxim & Gay Co. for money telegraphed by new customers. Some individual remittances were as high as $2,000. The money telegraphed us amounted to about $150,000, and within ten days eighty per cent of our own dividend checks were returned to us by our customers, endorsed back to us with instructions to double their bets, and within two weeks we were able to figure that in the neighborhood of $375,000 was sent us as a result.

The Rise and Fall of Maxim & Gay

A DISASTROUS NEWSPAPER WINDUP

DURING THE PROGRESS of the New Orleans meeting, I purchased a controlling interest in the New York *Daily America*—a newspaper patterned after the *Morning Telegraph*—from a group of members of the Metropolitan Turf Association, who had sunk about $75,000 in the enterprise. The *Morning Telegraph* was in the hands of a receiver. I calculated that, by transferring the Maxim & Gay advertisements from the *Morning Telegraph* to the *Daily America*, I could make the *Daily America* pay and force the *Morning Telegraph* out of the field. Later, the late William C. Whitney, who was a shining light on the turf as well as in finance, was induced to purchase the *Morning Telegraph*. Then trouble began to brew for me.

One morning I was summoned to the Nassau St. offices of August Belmont.

"For the good of the turf, you must omit your Maxim & Gay advertisements from the *Daily America* and other newspapers hereafter," declared Mr. Belmont on my entering his room.

"Why?" asked I.

"They flagrantly call attention to betting on the races," he repiled.

"But you allow betting at the tracks."

"Yes," he replied, "but public sentiment is beginning to be aroused against betting, and an attack is bound to result."

It occurred to me that at that very time Mr. Whitney was engaged in disposing of his stock in various traction enterprises in New York to Mr. Belmont and his syndicate, and that in all probability Mr. Whitney had sought the assistance of Mr. Belmont to put the *Daily America* out of business in this way. It was apparent that the *Daily America* would lose money fast without the Maxim & Gay advertising. Maxim & Gay, too, would practically be compelled to close up shop if it could not advertise. I promised to consider.

Returning to the *Daily America* office, I decided to pay no attention to Mr. Belmont's request, having become convinced that it was conceived in the interest of the *Morning Telegraph*.

A few days later I was again summoned over the 'phone to Mr. Belmont's office. When I was ushered into Mr. Belmont's presence he said, "If you don't quit advertising the Maxim & Gay Co. in the *Daily American*, I will see William Travers Jerome, and he will stop you."

Mr. Jerome was then District Attorney, and the idea of doing anything that Mr. Jerome considered illegal appalled me.

"If Mr. Jerome sends word to me that the Maxim & Gay advertising is illegal, I will discontinue it," I said. I did not hear from Mr. Jerome, and so went on with the advertising.

Within a few weeks the Washington race meeting opened at Bennings. When the Maxim & Gay staff reached there, we were all informed that the post office department was about to begin an investigation into our business affairs, and all of our staff voluntarily appeared before the inspectors and underwent an examination. Our books were also submitted. This investigation, coming on the heels of Mr. Belmont's threat, convinced me that the influence of Mr. Belmont and Mr. Whitney reached all the way to Washington, and I concluded that if I did not discontinue the Maxim & Gay advertising in the *Daily America*, and then, of course, discontinue the *Daily America*, they would make serious trouble. So I hung out the white flag. I announced my retirement from Maxim & Gay Co. and offered to sell my paper to Mr. Whitney.

My exchequer was low. Nearly every dollar I had made in the Maxim & Gay enterprise had been lost by me in plunging on the races myself.

During the following week Mr. Whitney received me at his palatial home on Fifth Avenue just after his breakfast hour. He interviewed me for about an hour, obtained my price on the paper, which was what I had put into it, namely $60,000, and promised to cable to Colonel Harvey, then, as now, the distinguished editor of the Harper publications, who was in Paris, asking his advice, saying that Colonel Harvey advised him in all newspaper matters. I did not hear from Mr. Whitney again; but I did discover that my business manager was in close communication with Mr. Whitney and that the state of my financial condition every evening was being religiously reported to him.

A few weeks later I was compelled to put the paper in the hands of a receiver, and a representative of Mr. Whitney bought it for $6,500, or about 10 cents on the dollar, and put it to sleep, leaving the field to the *Morning Telegraph*. From that moment the *Morning Telegraph*, which for a short period had been refusing all tipster advertising, resumed the acceptance of such business and has continued that policy up to this day.

A year after I retired from Maxim & Gay, Attorney-General Knox decided that racehorse tipping is an offense against the old lottery law, and those who now advertise tips instruct that no money be sent by mail.

Having lost the *Daily America* and having blown the Maxim & Gay Co., I was again broke. But my credit was good, particularly among racetrack bookmakers. That summer 1904, I became a racetrack plunger, first on borrowed money and then on my winnings. By June I had accumulated $100,000. In July I was nearly broke again. In August I was flush once more, having recouped to the extent of about $50,000. Early in September I went overboard; that is to say, I quit the track losing all the cash I had and owing about $6,000 to a friendly bookmaker.

Disgusted with myself, I longed for a change of atmosphere. I stayed around New York a few days, when the yearning to cut away from my moorings and to rid myself of the fever to gamble became overpowering. I bought a railroad ticket for California and, with $200 in my clothes, traveled to a ranch within fifty miles of San Francisco, where I hoed potatoes and did other manual labor calculated to cure race track-itis. In less than six weeks I felt myself a new man, and decided to stick to the simple life forever more —away from race tracks and other forms of gambling.

But I didn't.

It was in New York that George Graham Rice learned his entrepreneureal skills by operating a tipping service for horse racing, and in central Nevada he employed his promotional and advertising skills to great advantage.

San Francisco, 1904

2. Mining Finance at Goldfield

I HAD NEVER VISITED San Francisco. Being close to the city of the Golden Gate—within fifty miles—I decided to take a look. So one evening, in the late Fall of 1904, I packed my grip and within two hours was comfortably housed in the old Palace Hotel.

The first man I met on entering the lobby was W. J. Arkell, formerly one of the owners of *Frank Leslie's Weekly* and of *Judge*.

"Hello, Bill!" I exclaimed. "What are you doing here?"

"Same as you," he answered. "Morse trimmed me in American Ice, and I'm broke. I am in hock to the hotel. They think I am worth $2,000,000. I haven't 20 cents."

During the evening we consoled each other over a series of silver gin fizzes, several of which Arkell paid for with the stub of a pencil. My companion promulgated a scheme for the quick putting on their feet of two Eastern rovers adrift in the big Coast city, and that night there was formed the W. J. Arkell Advertising Agency. Then the horse tipping firm of "Jack Hornaday" was established. I declared that I preferred to have little to do with it except to show "Willie" how it had been done in New York by Maxim & Gay.

"I will do it for you, Bill," I said; "but no more for me—I've had enough."

"Jack Hornaday" advertisements appeared daily in all the San Francisco papers. Capable clockers and handicappers were hired and some excellent information was obtained. Racegoers got a run for their money.

But something happened. The racetrack trust, which enjoyed a big pull in the San Francisco *Examiner* office, soon realized that somebody outside of the inner circle was getting the public's money, and every day that "Jack Hornaday" tipped a loser the *Examiner* carried on its sporting page a notice to the effect that "Jack Hornaday's" tip resulted very disastrously to his clients.

Tonopah to Goldfield Stage 1905

Goldfield Roulette Game

A PARTNERSHIP OF PURE NERVE

"JACK HORNADAY" DISCONTINUED business. I began to like San Francisco and the Coast. Being thrown among Arkell's associates in the Palace Hotel lobby, from time to time I naturally heard a great deal of talk about the new Nevada mining camp of Tonopah.

"Rice," said Arkell one evening, "come with me up to Tonopah and be my press agent. We will get hold of a mining property up there, promote a company and make a barrel of money."

"What do you know about mines?" I asked.

"Well, I've lost enough in 'em to know a great deal," he answered.

"I don't know a mine from a hole in the ground, and I know nothing about the stock brokerage business; so I don't see how I can be of any assistance," I said.

"Don't let that bother you," he replied. "I'll show you how. You come with me."

"I will go on one condition," I said. "I am in for half on anything you do."

We shook hands and it was a bargain. We went to the depot. I had a trifle less than $150 in my pocket. Arkell had $75.

"Suppose we get stranded out there, what will happen?" I propounded.

"Oh, forget it!" he answered. "How can a couple of Easterners like us, wide awake and with phosphorus brains, get stranded in a place where they dig silver and gold out of the ground?"

We journeyed to Tonopah—a 36-hour ride. The altitude is 6,000 feet, and it was cold, nasty, penetrating winter weather. During the last hundred miles of our journey across the mountainous desert we looked out of the car window and saw trainload after trainload of what was said to be ore coming from the opposite direction, and we decided that Tonopah was a sure enough mining camp and that some of the sensational stories about bonanza mines that we had heard were really true.

"BUCKING THE TIGER" ON THE DESERT

ARRIVING IN TONOPAH after dusk, we sought hotel accommodations. The best we could get was a bed in a forbidding looking one story annex, walled with undressed pine and roofed with tarpaulin. It was located feet to the rear of the hotel, which was already crowded with miners and soldiers of fortune drawn from all quarters of the world by the mining excitement. Its aspect was so inhospitable that Arkell and I decided not to retire for a while. We gravitated out toward the barroom, where the click of the roulette wheel caught our ears.

We sat down to watch the game. Soon we were buying stacks of checks and ourselves bucking the tiger excitedly. In an hour the remnants of my $150 passed to the ownership of the man behind the game, and Arkell had put his last two bit piece on the black and lost.

I looked at him. He looked at me. "Umph!" he grunted. "Better hit the feathers!"

Meekly I followed him to the annex. When we got under the soiled gray woolen blankets, I remarked: "I've got a cane and an umbrella and three suits of clothes. Do you think we can sell them in the morning for enough to provide breakfast money?"

"Oh, come off!" exclaimed my partner. "Wait till I present my card around this burg in the morning; then we will get all the breakfast we want."

We awoke hungry, as all men have a habit of doing when they are broke.

"I am going over to the Montana-Tonopah Mining Co.'s office," said Arkell. "A mining engineer by the name of Malcom Macdonald makes his headquarters over there and he wants to sell some mining properties at Goldfield and in other parts of the state for about three million dollars."

"Three millions!" I exclaimed.

"Yes," said Arkell. "I'll get the facts and wire them to my friend Joe Hoadley in New York."

"Say, Bill," I remonstrated," they have a privately owned jerkline telegraph in this town, and if you send any phony telegrams over the wire, they'll be on to you. So don't do any of that kind of business."

"Nothing of the kind!" replied he promptly. "Any message I send to Hoadley he'll answer."

"I guess you have it fixed on the other end," I remarked. He laughed.

We strolled over to the State Bank and Trust Co. building, across the street, and there met Malcolm Macdonald, a mining engineer from Butte, Montana, and his friend, Mr. Dunlap, who was at the time secretary of the Montana-Tonopah Mining Co. The conversation was not more than five minutes old when Arkell suggested that he would like to eat breakfast, but "didn't want any restaurants in his," intimating that he would like to have some good, old fashioned home cooking. Mr. Dunlap remarked modestly that the camp was too young to boast of much home cooking, but that if we would be his guests he would guarantee to make arrangements for some special cooking at the Palace Restaurant.

BIDDING $3,000,000 WHEN BROKE

AFTER BREAKFAST, which consisted of mountain trout, the flavor of which was more delicious than anything I had tasted in many years—probably because of the artificial hunger which an empty purse had created—we returned to the office of the bank. There Arkell explained to Mr. Macdonald that he wanted a big mining proposition or nothing. He said he represented big Eastern capital and that he was prepared to pay from one to three millions for the right kind of property. Mr. Macdonald named some mines and prospects which he said he was willing to sacrifice for $3,000,000.

One of them was the Simmerone, of Goldfield, which Mr. Macdonald offered for $1,000,000. We afterward learned that he had paid $32,000 for it. At that time there was a six foot hole in the ground, and the whole property contained less than five acres. A stockade had been built around the workings on account of the extreme richness of the ore that had been opened at grass roots.

Mr. Macdonald also offered for sale a lead property at Reveille and a lead-silver property at Tybo, both situated about 70 to 100 miles from a railroad. (Later these properties, along with some others, were promoted by Charles Minzeshimer & Co., a New York Stock Exchange house, as the Nevada Smelters & Mines Co. and passed on to the public at a valuation of $5,000,000. The market value of the entire capitalization of this company is now less than $10,000.) These "mines" were to be put into the deal at $1,000,000 each.

MILLIONS IN THE VISTA HELD NO CHARMS

ARKELL WROTE A DISPATCH to the East in the presence of our newly made friends, describing the offering. Then he and I held a consultation, and he vouchsafed the information that we would certainly get a free automobile ride to Goldfield and have a chance to see there the new boom mining camp.

I got cold feet. Arkell's talk of visionary millions in that bleak environment of snow-clad desert and wind-swept mountain didn't enthuse me at all. I protested against the proposed trip to Goldfield, and insisted that I should be allowed to telegraph for money with which to return to the Coast.

But Arkell persisted. He declared that the expense of the trip to Goldfield and back to Tonopah would be borne by the vendors of the mines and that our return trip to San Francisco would be delayed only one day. I left my grip, umbrella and cane in Tonopah, intending to return the same evening, and boarded the automobile for Goldfield.

Arrived in Goldfield, we were escorted to the Simmerone. Arkell appeared to be very much impressed, although he remarked to me a few minutes later that he would not give $34 for the whole layout. And therein he was

A Saturnalia of Speculation at Goldfield

Goldfield's Mines and Streets Throng with Activity

wise. The Simmerone was later capitalized for 1,000,000 shares, each share of a par value of $1, ballooned on the San Francisco and Goldfield stock exchanges to $1.65 a share, and then allowed to recede to nothing bid, one cent per share asked. The rich ore petered out.

There was an indefinable something in the atmosphere of Goldfield—a new, budding mining camp, at an altitude of 5,000 feet and on the frontier—that stirred me, and I decided to stay awhile.

Arkell determined that he would go back to Tonopah and get an option on the control of a mining company known as the Tonopah Home, which Mr. Dunlap had mentioned to him in the automobile enroute to Goldfield. He said he would then go to San Francisco to promote it. The reason why he decided to handle the Tonopah Home, I afterward discovered, was that it was already incorporated and stock certificates had been printed, thereby eliminating the delay and expense incident to preparing something for the immediate consumption of the San Francisco public.

"How am I going to subsist here for a few days until I can begin to make a living?" I asked Arkell.

"How am I going to get back to Tonopah and from there to San Francisco?" Arkell asked me.

At that moment we stood in front of the Goldfield Bank and Trust Co.'s building—a tin bank literally as iron and tin. A few months later, when the bank went up the flume, the cash balance found in the safe aggregated 80 cents.

"You take me into this bank, introduce me and I will cash a check," he said.

"A check on what?" I asked.

"On my bank in Canajoharie, New York," he said. "I was born and brought up there, and they wouldn't let one of my checks go to protest. Besides, I can get back to 'Frisco and protect it by telegraph, if necessary, before it reaches Canajoharie."

We entered the bank. I introduced myself to the cashier as an Eastern newspaper man, and then introduced W. J. Arkell as the former publisher of *Leslie's Weekly, Judge,* and so on.

After a brief parley, Arkell exchanged his paper for real money to the amount of $50. On leaving the bank, I said:

"Now, Bill, come across! I'm flat broke, on the desert."

He handed me $15. I was satisfied, because he needed all of the $35 to get back to civilization.

HUMAN INTEREST VERSES TECHNICAL MINING

AFTER ARKELL'S DEPARTURE for Tonopah I went to the office of the Goldfield *News* and asked for a job. I got it, at $10 a day. My first assignment was to interview an old miner named Tom Jaggers. I wrote what I considered a first class human interest story, and handed it to the owner and editor, "Jimmy" O'Brien. He thought it was fair writing, but not the sort of matter the Goldfield *News* wanted. It wanted technical mining stuff. Of course I didn't know a winze from a windlass, nor a shaft from a stope, and some of the weird yarns I handed in about mine developments certainly did make Mr. O'Brien jump sideways. Within a week I was discharged for incompetency.

I was not at all appalled at losing my job on the Goldfield *News*. I had begun to like the life and was convinced there were some real gold mines in the camp. I was a tenderfoot and knew little or nothing about the mining business, but the visible aspect of shipment upon shipment of high-grade ore leaving the camp by mule team was convincing. What probably impressed me most was the evident sincerity of the trail blazers who had been on the ground since the day the camp was born. These men had suffered all kinds of hardships to hold their ground and make a go of the camp which, when discovered, was situated 100 miles from a railroad station and at least 25 miles from a known water supply. Tradition said that men had died of thirst on the very spot where Goldfield was now adding daily to the world's wealth.

My environment became an inspiration. There were a few penny mining stock brokerage firms doing business with the outside world, and the idea of starting an advertising agency appealed to me strongly. Here was an opportunity for the great American speculating public to "take a flyer" on something much more tangible and lasting than a horse race, I determined.

Failing to locate a furniture store I ordered a long, rough, pine board table made by a carpenter, rented desk room from the Goldfield Bank and Trust Co. right in front of the cashier's counter, and secured the services of an expert male stenographer from Cripple Creek. The Goldfield-Tonopah Advertising Agency was born.

BEGINNING THE ADVERTISING BUSINESS

THE IDEA OF APPLYING to the American Newspaper Publishers' Association for recognition did not occur to me. I did not know that such was the practise of agents. I did believe, however, from my ad writing experience with the Maxim & Gay Co. in New York that I could write money getting advertising copy. Further, my experience in making contracts with advertising agents for the publication of Maxim & Gay's advertising in the newspapers throughout the land had, it seemed, conveyed to me sufficient information regarding that end of the business to fortify me in my new field.

Next morning I entered the office of the Mims-Sutro Co., a newly established brokerage firm, and urged advertising.

"We are already spending about $100 a month," said the manager.

"One hundred dollars a month!" I exclaimed. "Why, you ought to be spending that much every hour!"

At first they thought me to be a fanatic on the subject, but within a fortnight I succeeded in inducing them to spend $1,000 in a single day for advertising. It was not, however, until after I had shown them how to follow up their correspondence successfully that they began to believe in me. I wired to nearly all of the important city newspapers throughout the country for rates. After obtaining their replies I decided to spend $500 in the Chicago Sunday *American*, and $500 in the San Francisco *Examiner* in one issue. I forwarded the copy with the money, and it appeared promptly. The results were good—so good, indeed, that within two months the Mims-Sutro Co. was spending at the rate of from $5,000 to $10,000 a week for advertising, and my commissions amounted to thousands.

My contracts with the advertisers required them to pay me one-time rates, and my contracts with the publishers permitted me to send in copy at long-time rates, and the profit was about 45 per cent. And inasmuch as I always sent cash with the order, my copy was in great demand. Indeed, my agency was fairly inundated day after day with blank contracts from newspapers all over the country, the managers of which were clamoring for the Goldfield business. In addition to the Mims-Sutro account, I soon had many others; in fact, I had all the others. Within six months after my arrival in Goldfield my agency netted me $65,000.

SOME ADVERTISING THAT PAID

MY SECOND BEST CUSTOMER was January Jones, the noted Welsh miner, and later, when the corporation of Patrick, Elliott & Camp swung into business as promoter, I placed its advertising. I held it, too, until the death of C.H. Elliott, when the control of that firm fell into other hands and it ultimately went out of business. In the course of three years my advertising agency inserted in the neighborhood of $1,000,000 worth of advertising in the newspapers of the United States, chiefly those of the big cities, and all of the advertising made money. It simply had to make money, because the brokers who did the advertising had little or nothing to begin operations with except the mines, and the mines were not their property.

The most remarkable feature of that advertising campaign to me was that I had never been a stock broker, had never been a mine promoter, and had

PROSPECTUS

of

Nevada Copper Butte Mining Co.

BUTTE MOUNTAIN

State of Nevada

The Goldfield Mascot Mining Co.

The Patrick Investment Co.
Fiscal Agents

402, 403, 404, 405
Empire Building
Denver, Colorado

PROSPECTUS
OF
The Red Fox Bullfrog Mining Company

Incorporated under the Laws of Arizona

Capital Stock 1,000,000 Shares
Par Value One Dollar Each
Full Paid and Non-Assessable

499,995 Shares in the Treasury

MINES SITUATED IN
BULLFROG DISTRICT, NEVADA

PRINCIPAL OFFICE
Suite 20, Toltec Block, Denver, Colo.

Officers:
L. L. PATRICK, President
W. F. PATRICK, Vice-President and General Manager
J. M. PATRICK, Secretary

Directors: WM. F. PATRICK
JAS. M. PATRICK

never been in a mining camp before; but still, despite my utter lack of knowledge, to begin with, of the technical end of the business, my advertisements pulled in the dollars.

I was an enthusiast. I believed in the merits of the camp, and my enthusiasm undoubtedly carried itself to the readers of my advertisements. But the quality of the advertising copy did not entirely explain my success in bringing the money into Goldfield. The stock offerings undoubtedly struck a popular chord. Tens of thousands of people who for years had been imbibing the daily financial chronicles of the newspapers, but whose incomes were not sufficient to permit them to indulge in stock market speculation in rails and industrials, found in cheap mining stocks the things they were looking for— an opportunity for those with limited capital to give full play to their gambling, or speculative, instinct.

Time and again promotions were almost completely subscribed by telegraph in advance of mail responses reaching Goldfield; and it frequently needed but the publication of a half page advertisement in 40 or 50 big city newspapers, of a Sunday, to bring to Goldfield by wire before Monday night sufficient reservations to guarantee oversubscription in a few days.

It was easy to give full play to my penchant for experimenting, in the evolution of mining stock promotion in Goldfield. The old system, and the one which recently has enjoyed much vogue among financial advertisers, was the endeavor first to get names of investors rather than immediate results from the advertisements, and to follow them up by correspondence. In spending the first $1,000 appropriated for advertising from Goldfield, I split the money between two newspapers on one day. I constructed large display advertisements and appealed for direct, quick replies. This succeeded.

BUILDING GOLD MINES WITH PUBLICITY

A LITTLE LATER I organized a news bureau as an adjunct of the advertising agency.

It is acknowledged that this news bureau accomplished much for Nevada. As a matter of fact, it is generally conceded by Goldfield pioneers and by mining stock brokers throughout the country that the news bureau was directly responsible for bringing into the State of Nevada tens of millions of dollars for investment, and was indirectly responsible for the opening up of the Mohawk and other great gold mines of the Goldfield camp and of the state.

The prospectors who located Goldfield were without means. George Wingfield, the man who is now president of the merged Goldfield Consolidated, came into the mining camp with only $150. No funds of consequence

Occupying center stage in stock promotion and manipulation were the Jumbo, Red Top, Laguna, and the Mohawk Mining Companies. In its first four years the Mohawk produced $8,832,000. These four companies and smaller properties later merged to form the famous Goldfield Consolidated which produced $50.4 million.

were available from home sources. The money that later made Goldfield the "greatest gold camp on earth" came from the outside, and the news bureau secured it by focusing the attention of the American public on the great speculative possibilities of investments in the mining securities and leases of the camp. One of the leases, known as the Hayes-Monnette, operated with Chicago money, afterward opened up the great Mohawk ore deposit at a period when there was no money in the treasury of the Mohawk Mining Co. to do its own development work. And there are scores of other instances which bear me out.

I was head of the news bureau, and the news bureau was Nevada's publicity agent. I have always considered my work in this direction in the light of an achievement. No one contributed a dollar to the news bureau except myself.

HAIR-RAISING STORIES FOR DISTANT READERS

THAT NEWS BUREAU, with its headquarters on the desert, at a time when water was commanding $4 a barrel in Goldfield and coal could not be obtained in the camp for love or money, was operated with as much calculating judgment as it could have been were it subsidized by the most powerful interests in America. Human interest stories that were written around the camp, its mines and its men, were turned out every day by competent newspaper men. These were forwarded to the daily newspapers in the big cities of the East and West for publication in the news columns.

Most of the stories were accepted and published. Whenever hesitancy was observed, publishers were tempted by the news bureau with large advertising copy to continue to give the camp publicity.

Of such great assistance in arousing public interest did I find this work that noted magazinists like James Hopper were imported to camp and pressed into service by the news bureau to write readable stories. At times, when public interest appeared to lag, the wires were used by the camp's newspaper correspondents to obtain publicity for all kinds of sensational happenings that were common on the desert. Reports of gold discoveries, high play at gambling tables, shooting affrays, gamblers' feuds, stampedes, holdups, narrow escapes, murders, and so forth, were used to rouse the public's attention to the fact that a mining camp called Goldfield was on the horizon.

I felt confident that the speculating public was going to make a great big killing in Goldfield. Tonopah, 26 miles to the north, was making good in a wonderful way. It had already enriched Philadelphia investors to the extent of millions. I could see no reason why Goldfield should not at least duplicate the history of Tonopah. Never in my life had I lived in an environment that inspirited me as this one. The visages of those around me were, as a rule, roughly hewn; the features of many were marked with all the blemishes that had been put upon them by time, by sleepless nights, by anxiety and by contact with the elements; but courage, sincerity and honesty of purpose were written in every line of their faces.

I became imbued with the idea that investors who put their money into Goldfield stocks were not only going to get an honest run for their money, in that the mines were going to be developed and many would make good, but that the opportunity for money making, if embraced by the public at that time, would earn a great reputation for the man who educated the public to a full understanding of the situation.

THE MERCURY OF SPECULATION

MINING STOCK SPECULATORS and investors at a distance who responded to the red-hot publicity campaign which marked those early days of Goldfield rolled up enormous profits, and I made no mistake. Terrific losses came eighteen months later, as a result of a madness of mining stock speculation which followed on the heels of the great Mohawk boom and the merger of various Goldfield producers into a $36,000,000 corporation. This was taken advantage of by wildcatters in every big city of the country, and the public was fleeced to a finish. But of this more and a plenty later.

In those early days my agency advertised Goldfield Laguna at 15 cents per share in order to finance the company for mine operations. Within a year thereafter Goldfield Laguna sold at $2 a share on the San Francisco Stock Exchange, and was absorbed by the Goldfield Consolidated at that figure. And there were many others which duplicated or exceeded the performance of Laguna.

At the time of which I tell, when Laguna was promoted at 15 cents, Goldfield was about a year old. A population of about 1,500 had gathered there from all sections of the country. There were mining experts from Salt Lake, San Francisco and Colorado, and miners from every part of the Western mining empire; saloon keepers from Alaska and Mexico; real estate brokers from practically every Western State and a scattering of "tin-horns." It was about as motley a gathering as one could find anywhere in the world, but compositely they were a sturdy lot.

The camp was enjoying its maiden boom. In sixty days real estate values had jumped from $25 for a lot on Main Street to $5,500. Roughly constructed business houses banked the main thoroughfare for two or three blocks. The heavy traffic incident to hauling in supplies from Tonopah had ground

Promoters like George Graham Rice did not end their day of promotion and banking at 5 p.m.

High-stake gaming continued far into the evening as captured by central Nevada photographer Ned Johnson, in Goldfield 1906-1907.

the dirt of the street into an impalpable mass of dust to the depth of fifteen inches, and the unchecked winds of the desert, sweeping from the Sierra Nevada to the high uplifts east of Goldfield, whipped the dust into blinding clouds that daily made life almost unendurable.

Practically the entire population was housed in tents that dotted the foothills. At night time these presented the appearance of an army encampment. Provisions were scarce and barely met the requirements. The principal eating place was the Mocha Café, which consisted of a 14 by 18 foot tent with an earthen floor and a roughly constructed lunch counter. Here men stood in line for hours, waiting to pay a dollar for a dirty cup of coffee, a small piece of salty ham and two eggs that had long survived the hens that laid them.

The popular rendezvous was the Northern saloon and gambling house, owned and managed by "Tex" Rickard and associates. Here fully seventy-five per cent of the camp's male population gathered nightly and played faro, roulette and stud poker, talked mines and mining, sold properties, and shielded themselves from the blasts that came with piercing intensity from the snow-capped peaks of the Sierra. The brokers of the camp gathered every night in the Northern and held informal sessions, frequently trading to the extent of 30,000 or 40,000 shares of the more active stocks.

The mining stocks which were advertised through my agency in those early Goldfield days were generally of the 10, 20, and 30 cent per share variety. The incorporators of the companies were enthusiastic on the point of their prospect making good, but I argued to myself that if the chances of any mining prospect of this character proving to be a mine were only about one in 25 or one in 50, and my agency advertised 25 or 50 companies of the average quality, and one of them made good in a handsome way, he who purchased an equal number of shares in each would at least break even with the profits from the one winner.

Later this principle was "knocked into a cocked hat" for conservatism by Mohawk of Goldfield advancing from 10 cents to $20 a share, proving that if Mohawk had been one among 50 companies, the shares of which were purchased by an investor at 10 cents, he would have gained handsomely. Early purchasers of Mohawk gathered 200 to 1 for their money, many times more than could usually be won on a long shot at the horse races, and not so very much less than was formerly won by lucky prize winners in the Louisiana Lottery. And Mohawk was only one of a dozen of the early ones which advanced in price on the exchanges and curb markets more than 1,000 per cent.

At this early stage in Goldfield, wildcatting was not indulged in from the camp, unless this long-shot gambling in shares of prospects can by a grave stretch of imagination be termed such, the promoter-brokers being able to offer stocks of close-in properties. Among the prizes were Red Top, which advanced within two years thereafter from eight cents to $5.50 per share; Daisy, which sky rocketed from 10 cents to $6; Goldfield Mining, which soared from 10 cents to $2; Jumbo, which improved from 50 cents to $5; Jumbo Extension, which rose from 15 cents to above $3; Great Bend, which jumped from 20 cents to around $2.50; Silver Pick, which moved up from 10 cents to $2.65; Atlanta, which was promoted at 10 and 15 cents and sold up to $1.25; Kewanas, which was lifted from 25 cents to $2.25, and others. Wildcatting in a small way was prosecuted in Goldfield's fair name even in those days, with Denver as the headquarters of the swindlers.

Eighteen months later, when the Mohawk mine of Goldfield was in the midst of its greatest half year of production, at the rate of $1,000,000 a month, and the consolidation of the important mining companies of the camp was in progress, wildcatting became general from office buildings in the large cities. There were more than 2,000 companies incorporated during this last period, not one of which made good, and the public lost from $150,000,000 to $200,000,000 as the result of this operation alone. Fully $150,000,000 more was lost by the balooning to levels unwarranted by mine showings of listed Goldfield stocks on the New York Curb and the San Francisco Stock Exchange, at the same time.

But I am ahead of my story.

It was late in the spring of 1905. I had been at work in Goldfield more than six months, and my campaign of publicity was beginning to gather momentum. The mines, however, were not at the moment keeping lively pace. The Mohawk was yet undiscovered.

THE BIRTH OF BULLFROG

AT THIS JUNCTURE the new mining camp of Bullfrog, 65 miles south of Goldfield, was born. My publicity facilities were sought by owners of properties in Bullfrog to put the camp on the map.

C. H. Elliott, a Goldfield pioneer, put an automobile at the disposal of myself and my stenographer, and we departed for Bullfrog. Elliott and his associates had staked out a townsite which they called Rhyolite. I was presented with seven corner lots on my arrival, to help along my enthusiasm.

There, on the saloon floor of a gambling house, which was the chief place of resort in the camp, I met for the first time George Wingfield, then the principal owner of the Tonopah Club at Tonopah, a gambling house which had lifted him from the impecunious tin-horn gambler class to the millionaire

THE FIRST YEAR'S RECORD OF
BULLFROG NEVADA
AND ITS MARVELOUS MINES

First discovery—the Bullfrog—Aug. 9, 1904.

First shipment of seven tons from first discovery netted $843.90 per ton.

First shipment from the second great discovery—the Shoshone—netted $2,300.00 per ton. The last shipment of 30 tons netted $1,150.00 per ton, with thousands of tons to follow.

A year ago the Bullfrog district was an unknown, unconquered desert waste; today it boasts of three lively, up-to-date towns—Rhyolite, Bullfrog and Beatty—with a combined population of 6,000, and increasing at the rate of 100 per day.

Surveying, grading and tracklaying are under way between Bullfrog and Tonopah, 90 miles northwest, and Las Vegas, 120 miles southeast, which will give this district easy and cheap connections by rail with San Francisco, Salt Lake City, Los Angeles and other large trade centers, and afford it the same facilities enjoyed by older camps. Before another year passes, Bullfrog will see the completion of both of these railroads.

A GOOD FIELD FOR INVESTMENT.

In addition to the large inflow of capital into the Bullfrog district for mining investments, much money has been already expended in many collateral enterprises which seem to promise equally as large financial returns as its bonanza mines. With the completion of the Bullfrog Water, Light and Power Company's

Jumbo Mining Company
OF GOLDFIELD

Pay to

Jumbo, Nev., Oct 24 1906. No. 483

John S. Cook & Co., Bankers
GOLDFIELD, NEVADA.

or order $4⁰⁰/₁₀₀

Dollars

R. Holt
TREASURER

JUMBO MINING COMPANY OF GOLDFIELD

PRESIDENT

division; United States Senator George S. Nixon, his partner; T. L. Oddie, later elected Governor of Nevada; Sherwood Aldrich, now one of the principal owners of the Chino and Ray Consolidated mines, and worth millions, and others who have since accumulated great riches.

[On the death of Mr. Nixon in Washington, D.C., in June 1912, Mr. Wingfield was appointed his successor as U.S. Senator by Governor Oddie. Mr. Wingfield's Goldfield newspaper felicitated its owner and pronounced the appointment to be logical and deserved. Mr. Wingfield, however, after hearing from Washington as to the manner in which the news of his appointment was received by members of the Senate, notified Governor Oddie three weeks later that he must decline the honor. He gave other reasons.]

They were on the ground and buying properties. Mr. Aldrich purchased the controlling interest in the Tramps Consolidated for about $150,000. It was incorporated for 2,000,000 shares of a par value of $1 each, a year later boomed to $3 a share in the New York Curb, and is now selling at three cents, without ever having paid a dividend.

Mr. Elliott had a large stock interest in the Amethyst mine and the National Bank mine, which were capitalized for 1,000,000 shares respectively, and he presented me with 10,000 shares of stock in each. He and his partner sold the control of the Amethyst to Malcolm Macdonald of Tonopah. Later, when Amethyst's neighbor, Montgomery-Shoshone, was selling at $20 per share, the market price of Amethyst was pushed up to above $1 a share on the San Francisco Stock Exchange, and I took my profit. The Amethyst has since turned out to be a rank mining failure, as has practically every other property in the camp, not one ever having earned a dividend.

The Bullfrog National Bank stock, representing another property that looked for a while as if it would make good, I disposed of on the San Francisco Stock Exchange at 40 cents a share, and I sold the town lots at figures which netted me, in all, in excess of $20,000 for my one day's trip to Bullfrog.

During my stay in Bullfrog I became very much impressed with the Montgomery-Shoshone mine. This property in fact was the powerful magnet which attracted everybody to the camp.

I was escorted through a tunnel seventy feet long. On each side as I walked were walls of talc. I was told these assayed in places anywhere from $200 to $2,000 a ton. Information was also forthcoming that the width of the ore body was more than seventy feet. (It afterward turned out that the tunnel had been run along the ore body and not across it, and that the ore body was about 10 feet wide.) Some specimen ore was given me to assay, and the returns were staggering, running all the way from $500 to $2,500 a ton.

In my enthusiasm I wrote stories about the property for publication which must have induced the reader to believe that when all the riches of that great treasure house were mined, gold would be demonetized. As a matter of fact, the stories from my news bureau, picturing the riches of that Golconda are said to have been indirectly responsible for the purchase of control of the property by Charles M. Schwab and his associates.

The history of the Montgomery-Shoshone is mournful but highly instructive. For purposes of exposition of pitfalls in mining stock speculation it possesses striking qualifications. Here are the facts:

Malcolm Macdonald, mining engineer, acquired a half interest in the mine from Tom Edwards, a Tonopah merchant, for $100,000, on time payments. On the strength of the showing in the 70-foot tunnel an effort was made to sell the control to the Tonopah Mining Co. at a profit. It did not succeed. Oscar Adams Turner, of New York and Baltimore, the promoter of the highly successful Tonopah Mining Co., which to date has paid back to the original stockholders $16 for every $1 invested, examined the Montgomery-Shoshone, and turned it down because the property did not show him any well defined veins or other marks of permanency, and the ore body appeared to him to be only a superficial deposit of no great extent.

Many a good prospect has been condemned by mining men of the highest standing, and has afterwards made good, particularly in Nevada. Mr. Turner's turn-down did not daunt the owners.

ENTER, CHARLES M. SCHWAB

ENGINEER MACDONALD INCORPORATED a company for 1,250,000 shares of the par value of $1 each, to own and operate the mine. Investors were permitted by him to subscribe for small blocks of treasury stock at $2 per share. Shortly afterward Mr. Macdonald and the owner of the other half interest, Bob Montgomery, sold a controlling interest to Mr. Schwab and associates for a sum which has never been made public. Mr. Schwab at once reorganized the company, took in two adjoining properties that were undeveloped, and changed the capitalization to 500,000 shares of the par value of $5 each. He, in turn, permitted his friends and the public to subscribe for the new stock at $15 per share. Later the shares advanced to $22 on the New York Curb.

Undoubtedly Mr. Schwab thought well of the proposition, for he loaned the company $500,000 to build a reduction works on the ground.

To date the mine has failed to pay for its equipment. Work on the property has been abandoned and the mill has been advertised for sale.

The company still owes Mr. Schwab about $225,000, the net profits on the ore in six years being insufficient to repay his loan to the company. In fact, the enterprise has proved to be one of the sorriest failures in Nevada. The mine in six years produced $2,000,000 gross, and although mine and mill were operated in an economical way, the net proceeds from the ores were insufficient to pay off the Schwab debt. Recently the shares have been nominally quoted at from two to five cents on the New York Curb. The public's loss mounts into millions.

Investigation proves to me that Mr. Schwab was merely a mining come-on and allowed his enthusiasm to run away with him, but the public suffered just as much as if Mr. Schwab had perpetrated a cold-blooded swindle.

I have heard the question propounded by a stockholder, "What possible excuse could a man, with a good business head like that of Mr. Schwab, have for promoting the Montgomery-Shoshone at a valuation of $15 a share, or $7,500,000 for the property, afterward allowing the stock to be quoted up to $22 a share on the New York Curb, or at a valuation of $11,000,000 for the property, when, as a result of six years of mine operations, the company is practically insolvent?"

An excuse acceptable to mining men might be offered were the Montgomery-Shoshone property situated in a nest of other great mines, intrinsically worth many times the valuation placed on the Montgomery-Shoshone at the time of its promotion. Prospects of this variety, according to approved mining experience, are sometimes entitled to appraisement of great prospective value when neighboring mines have demonstrated deep-seated enrichment. But there was no such excuse in this case, because the deepest hole in the ground in the entire camp was less than 200 feet at the time the Montgomery-Shoshone was promoted by Mr. Schwab, and there was not a proved mine in or near the camp.

I was present in Reno about three years ago when Mr. Schwab passed through the divorce city enroute to California. At that time Montgomery-Shoshone had already cracked in price to around $3 a share, and stories were being published in Nevada that Mr. Schwab had been snubbed by members of an exclusive Pittsburgh club for recommending Montgomery-Shoshone for investment. Mr. Schwab, in hurriedly discussing the matter at the railroad station, was quoted to the effect that the property had been grossly misrepresented to him. This statement was widely published in Nevada. Thereupon, Don Gillies, Mr. Schwab's engineer in Nevada, who, with Malcolm Macdonald, was believed to be Mr. Schwab's mining adviser, telegraphed Mr. Schwab and asked point-blank whether he referred to him. Mr. Schwab answered that he did not. This denial was also given wide publicity. There was only one reasonable corollary, then, and that was that Mr. Schwab referred to Mr. Macdonald.

In fine, it appears that Mr. Schwab may have actually purchased the Montgomery-Shoshone on the sole representations of the vendor, the interested party, and may have actually promoted the property on the strength of the unverified representations of the vendor. It might be that the vendor did not misrepresent at all; he may have been too enthusiastic only, and communicated his enthusiasm to Mr. Schwab.

Possibly Mr. Schwab relied on newspaper accounts, and promoted the property on the strength of them. A letter from Mr. Schwab, which appears farther on, lends some color to this idea.

Even before this time Mr. Schwab had been in the mining game at Tonopah. His Tonopah venture was the Tonopah Extension. The control of the Tonopah Extension Mining Co. was bought by John McKane, later a member of the English House of Commons, from Thomas Lockhart at 15 cents per share. The capitalization was 1,000,000 shares. John McKane interested Robert C. Hall, a member of the Pittsburgh Stock Exchange, in the proposition. He, in turn, made a deal with Mr. Schwab. The stock then skyrocketed to above $17 a share on the San Francisco and Pittsburgh stock exchanges and the New York Curb. Afterward the price was allowed to recede to around 65 cents per share. During the past half year it has maintained an average quotation of $2.00 per share.

Although the market price of the shares at the time Mr. Schwab was believed to own the control was allowed to be advanced to a valuation for the mine of $17,000,000, the company has since failed to pay as much as $1,000,000 in dividends, and a quite recent appraisement by Henry Krumb, a noted engineer, of the net value of the ore in sight in the mine did not place it at so much as $1,000,000. The accuracy of this report is disputed, on the ground that the ore exposures at the time did not permit of fair sampling. This allows for a discrepancy, but hardly of $16,000,000.

After Tonopah Extension declined from around $17 a share to below $1.00 a share, it was alleged by Tonopah stockholders that Mr. Schwab and his associates had unloaded at the top. Mr. Schwab replied that he owned just as much stock after the market collapse as he did when he went into the enterprise. This was met with an allegation by some stockholders that while Mr. Schwab could probably prove that his interest was as large at the later period as it had been at the outset, it did not mean that Mr. Schwab and his *confrères* had not unloaded at the top and bought back at the bottom.

The following letter from Mr. Schwab to Sam C. Dunham, formerly U.S.

Census Commissioner to Alaska, afterward editor of the Tonopah *Miner*, and later mining editor of the *Mining Financial News* of New York when I was managing editor, denies personal guilt, although it leaves the reader free to believe that if Mr. Schwab, personally did not unload his stock at high prices, his associates might have done so.

CHARLES M. SCHWAB
111 BROADWAY
NEW YORK

Mr. Sam C. Dunham,
   Editor *The Tonopah Miner*,
    Tonopah, Nevada,

My Dear Mr. Dunham:

My attention has been called to your issue of Saturday, October 26, 1907. To such criticisms as that issue contained of me, I generally do not reply, as it is useless and only leads to further discussion. But your paper heretofore has been so uniformly kind to me, so fair in every respect, and as I have always regarded you as a friend, our relations having been so pleasant, it makes me feel that I would like to make a short reply to the criticisms mentioned, as showing the consistency of my position.

The only thing I criticised about Nevada was the inaccuracy of statements emanating from Nevada. You seem to attack me because of this statement, and the strength of my position is fully confirmed by your article because little, if anything, stated therein is true or accurate.

I will take up your statements one by one. You say I bought from John McKane $25,000 worth of stock of the Tonopah Extension Mining Company at 15 cents per share. This is absolutely untrue.

You say I bought 100,000 shares of Extension stock from Robert C. Hall at $6 per share, and paid for this stock with paper mill stock. No single part of that statement is correct. I never gave Mr. Hall any paper mill stock, nor did I buy 100,000 shares from him. The amount purchased from him was 60,000 shares. The price you state I paid him for the stock is not correct, and, as I stated, I gave no paper mill stock in exchange.

You say further that at the last annual meeting of Tonopah Extension stockholders, held in Pittsburgh last May, it developed that I had disposed of all the stock I purchased from Mr. Hall and over two-thirds of my original holdings of 166,000 shares. This is absolutely untrue. I am holding today exactly the amount I held after all purchases were made by me, and from the beginning, aggregating some 285,000 shares, and I think if you take the trouble to look up the records you will find my statement in this connection to be true. When I originally bought Extension there was also some stock in my name belonging to others, which I subsequently transferred to them, leaving my own holdings of 285,000 shares where they now remain intact in my personal possession.

Going on down the article, you say that I purchased control of Shoshone and Polaris for less than $2 per share. This statement of yours is inaccurate. You say I sold a large block of Shoshone stock at $20 per share. This is also without any truth whatever. The fact is that 3,000 shares were sold at this figure, $20, and these 3,000 came from the treasury of the company, all of which you will find a matter of record.

It is true that I have loaned the company nearly $500,000 to build the new mill, and I shall be glad to have any other stockholder in the company assume his pro rata share of this amount.

You wonder why I criticise statements from Nevada.

Respectfully yours,
(*Signed*) C. M. SCHWAB

The general impression in Nevada, as I have gathered it, is that Mr. Schwab's mining enterprises have been great disappointments to him, but that he did not lose any very large sum of money, and that the public did. His enemies go so far as to allege that he, his brother, and his brother-in-law, Dr. M. R. Ward, made millions out of the public.

I have an opinion, and I may be allowed to express it. Mr. Schwab, at the time he became a promoter of Nevada mines, was an expert steelmaker. He knew little or nothing about silver, gold and copper mines. The fact that friends in Philadelphia, who knew as little about the game as he did, had made a fortune in Tonopah (on the advice of a man who did know) should not have influenced him. Because the Mizpah mine at Tonopah, promoted by Oscar A. Turner as the Tonopah Mining Co., had made good in a phenomenal way, Pennsylvania stockholders had rolled up fabulous profits in the venture. Under this hypnotism Mr. Schwab fell for Tonopah Extension.

Later, when Tonopah Extension showed a market enhancement of more than $16,000,000, Mr. Schwab was in an ideal frame of mind to succumb to Montgomery-Shoshone.

And when Montgomery-Shoshone in the Bullfrog boom showed a market enhancement of $8,000,000, it did not take much argument to get him into Greenwater, another "bloomer," which is described further on.

Market profits were evidently alluring to Mr. Schwab. He failed to realize that his own great name was in large measure responsible for the rise in price of his securities.

Sam C. Dunham has informed me that Mr. Schwab told him he refunded to his personal friends in Pittsburgh, who subscribed for Montgomery-Shoshone stock on his recommendation, between $2,000,000 and $3,000,000. This ought to be convincing that Mr. Schwab was guiltless of any intent to profit at the expense of others.

Mr. Schwab's lack of caution, however, is instructive to the losing speculator. It furnishes a startling example of the danger in banking alone on an honored name for the success of an enterprise, and it also drives home the truth of the adage, "Every shoemaker should stick to his last."

Incidentally, Mr. Schwab's mining career points another moral. It is this: Don't think, Mr. Speculator, because a promoter represents the chances of profit making in a mining enterprise to be enormous, and you later find his expectations are not realized, that the promoter is *ipso facto* a crook. Big financiers are apt to make mistakes and so are little ones. Undoubtedly grave misrepresentations are made every day, and insidious methods are used to beguile you into forming a higher opinion regarding the merit of various securities than is warranted by the facts. But mine promoters are only human, and honest ones not infrequently are carried away by their own enthusiasm and themselves lose their all in the same venture in which they induce you to participate.

WHY THE BOTTOM FELL OUT

WHEN MONTGOMERY-SHOSHONE was enjoying its market heyday the Bullfrog Gold Bar Mining Company was promoted at around 15 cents a share on the usual million-share capitalization. A year later the price jumped to $2.65 on the San Francisco Stock Exchange, and the stock was widely distributed among investors. Recently the company was in the sheriff's hands. The biggest losers in this venture were Alabama people, who had great confidence in the promoters.

Other Bullfrog derelicts in which the public lost vast sums of money were Gibraltar, Bullfrog Steinway, Shoshone National Bank, Bullfrog Homestake, Bullfrog Extension, Denver Rush Extension, Mayflower, Four Aces, Golden Scepter, Montgomery Mountain, Original Bullfrog, etc., etc.

Mining stock brokers of the cities went into ecstasies over Bullfrog during the height of the boom in that camp. Philadelphia mining stock brokers fed Tramps Consolidated of Bullfrog to their clients. Pittsburgh brokers recommended Montgomery-Shoshone. Butte brokers placed large blocks of Amethyst. Gold Bar was distributed by brokers of the South. New York brokers were behind Gibraltar, Four Aces, Denver Rush, Montgomery Mountain, Eclipse, Golden Scepter, National Bank and a score of others. Practically every dollar of the millions invested in Bullfrog stocks has been lost.

The cause of the failure of the Bullfrog district to make good was not the absence of gold-bearing rock, for there is much of it in the district, but it has been found that the per ton values are too low to make the mines a commercial success. Bullfrog is situated on the desert and has no timber and but very little water. Promoters and investors did not realize this until mills were constructed. Then it was too late. If the camp were situated on the timbered shores of the Hudson River, the stocks of many of the mines of the district would probably be in great demand at above par.

Probably the most remarkable fact regarding Bullfrog is that its securities were more strongly recommended by Eastern brokers than the Goldfield issues and became more fashionable at this early period in Goldfield's history. Eastern brokers then had little confidence in Goldfield; and at the very time when the stocks of Goldfield representing inside properties, which later made good in an extraordinary way, were being offered, they advised their customers not to buy. The general cry then was that it was a fly-by-night offshoot of the first great Tonopah boom, and the idea prevailed in the East, because of the ascending influence of George Wingfield, then principal owner of Tonopah's leading gambling hall, that Goldfield was a haven for gambler's and wildcatters.

It was during the early days of the Bullfrog boom that my friend W. J. Arkell's career as a mining promoter came to a sudden end. It will be remembered that when he left Goldfield to go to Tonopah to make the Tonopah Home deal his cash capital was $35. He closed the transaction for the option on the million shares of Tonopah Home's capitalization at a price around five cents a share. Then our "partnership," of three days' duration, came to an end. Arkell journeyed back to San Francisco and there declared me out.

Arkell was a prominent figure for a while as a San Francisco mining stock

The vein is shown at a 45 degree angle at one of Manhattan's promotional mines

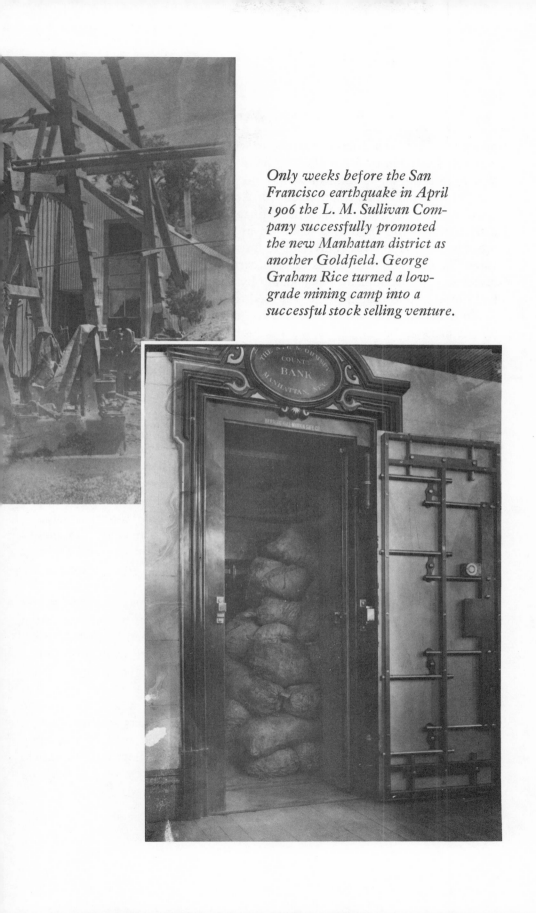

Only weeks before the San Francisco earthquake in April 1906 the L. M. Sullivan Company successfully promoted the new Manhattan district as another Goldfield. George Graham Rice turned a low-grade mining camp into a successful stock selling venture.

promoter. He listed Tonopah Home on the San Francisco Stock Exchange. Then he started in to sky rocket the price. The rise continued until the stock sold at 38 cents, an advance of about 700 per cent in a few months.

Then the psychological moment for Arkell arrived.

It leaked out that he had been financing his own stock on one-third margin and at the same time selling it, in like quantity, for all cash through other brokers. This was equivalent to borrowing $66\frac{2}{3}$ per cent of the market value. The brokers and banks did the carrying. When Arkell's tactics were discovered, indiscriminate short-selling by market sharp shooters ensued. Arkell's own hypothecated stock was used to make deliveries.

In order to hold his ground and to get the floating supply of the stock off the market, Arkell engineered a consolidation. The Tonopah Home Consolidated was incorporated, and holders of Tonopah Home stock were invited to exchange their original certificates for shares in the consolidated company.

Just then somebody threw a brick. The names of United States Senator George S. Nixon and Hon. T. L. Oddie, later Governor of Nevada, had been published as directors of the new company, and when these gentlemen saw the half page advertisements in which their names were used, and were informed that Arkell appeared to be on the ragged edge, they telegraphed to the San Francisco Stock Exchange denying connection.

Tonopah Home broke wildly on the announcement in the Exchange to something like three cents a share. Then it dropped to nothing. Arkell's methods were too raw, and I knew the smash had to come, sooner or later.

'Twas late in October 1905. Bullfrog was still in its heyday. Goldfield's initial boom seemed to be flickering. Work was going on day and night in the mines, but for want of fresh discoveries the camp was being deserted by some of the late-comers.

Out of town newspaper correspondents came upon the scene, and stories and pictures of the camp, labeled "A Busted Mining Camp Boom," etc., soon appeared in the Los Angeles and San Francisco newspapers. Goldfield mine owners were accused of beguiling the public. Promoters were gibbeted as common bunco men. Peculiarly enough, Bullfrog, younger sister of Goldfield, which has since proved to be such a graveyard of mining hopes, was immune. There men of substance were in control, the writers said, while Goldfield was portrayed as a stamping ground for gamblers and wildcatters. The stories had their effect even in Goldfield. Leading men of the camp began looking about for new fields to conquer. The majority of Goldfield mine owners had not fallen for Bullfrog, but the success of the Bullfrog stock company promotions in the East inspired them.

Behind the parades and the glib talk of the promoters, serious operators brought business-like efficiency to the camp of Manhattan

The great mining camp boom of Manhattan, 80 miles north of Goldfield, which followed, owes much of its success to these fortuitous circumstances. I was one of the first to get the Manhattan fever.

W. F. "Billy" Bond, a Goldfield broker-promoter whose ear was always glued to the ground, showed me a specimen of ore literally plastered with free gold. He said it came from Manhattan and that Manhattan was another Cripple Creek. It was only the night before that I had lost a good many thousand dollars "bucking the tiger." Faro was the pastime of practically everybody in Goldfield in those days, and I played for want of some other means of recreation and lost heavily.

I was as broke as the day I entered the camp. I bought blankets, a suit of canvas clothes lined with sheepskin, and a folding iron cot, all on credit. I packed the outfit off to Tonopah. There I climbed aboard an old, rickety stage coach of the regulation Far-Western type, and started for Manhattan. We rode over a snow-clad desert, up mountains and down canyons—a perilous journey that I would not care to duplicate. The $10 I had in my pocket, after paying my fare, was borrowed money. When I arrived that night at Manhattan, situated in a canyon at an altitude of 7,000 feet, I set up my cot on the snow, wrapped myself in my blankets and slept in the open. There were only three huts and less than a score of tents in the camp.

The next morning I strayed through the diggings. Sacks of ore in which gold was visible to the naked eye were piled high on every side. The Stray Dog, the Jumping Jack and the Dexter were the three principal producers. They honeycombed one another. I questioned some of the prospectors as to the names of the single claims adjoining the Stray Dog, Jumping Jack and Dexter. They informed me that there was one group of claims adjoining that could be bought for $5,000. With $10 in my pocket I proceeded to purchase it. I gave a check for $100, signed a contract to pay the balance of $5,000 in 30 days or forfeit the $100, and immediately started back to Goldfield to induce the president of the bank to honor my check on presentation. He did.

When I returned to Goldfield I carried with me many specimens of high-grade ore. They were placed on display in a jewelry store. There was great excitement, and before night a stampede from Goldfield to Manhattan ensued which in magnitude surpassed the first Goldfield rush.

A few days later I returned to Manhattan and sold my option for $20,000 cash. While I was there I met C. H. Elliott. Mr. Elliott had cleaned up in Bullfrog. He told me that he had formed a corporation partnership in Goldfield with L. L. Patrick, one of the owners of the great Combination mine—which was later sold to the Goldfield Consolidated for $4,000,000—and Sol.

Camp, a mining engineer from Colorado. The name of the concern was Patrick, Elliott & Camp, Inc. It was organized to promote mining companies. Mr. Patrick is now president of the First National Bank of Goldfield.

Mr. Elliott asked me to stay in camp for another day until he could pick up a good property. He made a deal with some cowboys for a large acreage embracing the April Fool group of claims, scene of the original gold discovery. Twenty leases on this property were in operation, and the surface showings were promising. If the ore went down, the mine would prove to be a bonanza. Mr. Elliott incorporated a company known as the Seyler-Humphrey to own and operate the ground.

We returned to Goldfield. My publicity bureau telegraphed the news of the Manhattan discoveries to a long chain of newspapers East and West. Then I put out a big line of display advertisements in the big cities, offering for sale stock of the Seyler-Humphrey. The entire issue of 1,000,000 shares of Seyler-Humphrey was oversubscribed at 25 cents a share within two weeks. This was the result of $15,000 worth of advertising, and the profits of the firm were $100,000. In quick succession Mr. Elliott promoted the Manhattan Combination and the Manhattan Buffalo. Within six weeks the firm's promotion profits amounted to approximately $250,000.

HOW ABOUT THE PUBLIC'S CHANCES?

I ASKED MR. ELLIOTT one evening, shortly after Patrick, Elliott & Camp earned their first $250,000 from their three Manhattan promotions, whether he did not think the public was entitled to subscribe for this stock at a lower price and at a smaller profit to his corporation.

I recall that he said: "The article we sell is something that somebody wants and is willing to pay for. What we have sold them is worth what we have charged. The fact that we are on the ground and have endured hardships entitles us to a good profit, provided the gold showings on the surface of the properties are not exaggerated. The sale of the stocks has been accelerated by your gift of presentation through advertisements. Big department stores and advertising specialists in the cities pay from $15,000 to $30,000 a year for that kind of talent, and we on the desert also have a right to avail ourselves of it."

"But suppose the properties don't make good?" I queried.

He answered: "It is not a case of excessive optimism for one to expect that Manhattan properties will make into mines, in the presence of such wonderful surface showings; and so long as we are not knowingly guilty of deception, no harm is done. If the Manhattan stocks we have promoted make

The Story of the L. M. Sullivan Trust

Every one in the United States of school age has heard of the "Days of the Comstock." The excitement of the palmy days upon the famous Mother Lode of the Comstock is being repeated in the world's greatest gold camp, Goldfield, and this fall more fortunes have been made in Nevada mining stocks than for any like period since the early '80's. And while fortunes have already been made and are being made daily, the Fickle Dame will continue to smile upon those who invest right in Southern Nevada stocks for some time to come. The great mining camps of the greatest mining State in the Union are in their swaddling clothes, as it were, and the developments of the next month, the next week—tomorrow—may make millionaires of those who fortunately took advantage of the opportunity and secured stocks in desirable properties.

There could be no better evidence of the confidence with which traffic in mining stocks of Nevada's camps is regarded by the investors of this State and the outside world than the fact that the considerable number of brokers making these stocks a specialty are reaping a rich harvest for themselves and their clients. Capital does not seek unprofitable investment, and brokers who have not that to offer which will yield a bountiful return must seek another field of enterprise. Not so the brokers and promoters of Goldfield, Tonopah, Greenwater and other of the State's camps. Their beginnings, as a rule, have been modest in the extreme, but the growth of their business has kept pace with the marvelous development of the districts in which they operate, and the constantly increasing demand for Nevada stocks, from the Greenwater district on the south to Fairview on the north, and the men, firms and corporations which have offices in the mining centers, enjoy, without exception, a prosperous and expanding business, the measure of success being determined only by the acumen and astute business perception which must affect results in every field of enterprise. In view of the above, it is indicative of the existence of business capacity of a high order in the individual members of the above corporation that the L. M. Sullivan Trust Company have attained to the distinction of occupying a place in the front rank of the brokerage and promotion firms, not only of the State, but the entire West as well.

The brokerage and promotion business of this largest and most successful house was established less than a year ago by L. M. Sullivan, president of the company. An extensive experience covering many years in the West proved invaluable to Mr. Sullivan in launching his business, an enterprise which has done more than any other one agency to make Nevada's great mineral wealth known to the world. The first thing he did was to secure the professional services of Mr. John D. Campbell, an eminent mining engineer, who is an acknowledged authority on mines and mining. Mr. Campbell was connected as consulting engineer with all the early successes of the L. M. Sullivan Trust Company of Goldfield, and John McKane's investments, as well as the extensive investments of Chas. M. Schwab, the steel magnate, and no mining property is taken on by the L. M. Sullivan Trust Company without an exhaustive personal examination and the endorsement of Mr. Campbell.

...Company

GOLDFIELD

SULLIVAN STOCKS HAVE RISEN IN VALUE One Million Dollars

Present Listed Market Value of Sullivan Issues............$2,900,000
Promotion Price of Sullivan Issues 1,900,000

Our Record

Lou Dillon Goldfield
SIXTY-SIX PER CENT PROFIT IN TWO WEEKS

Promoted by us last month at 30 cents per share. Now listed on the San Francisco and Salt Lake Exchanges and selling at above 50 cents. Good for $1 per share before January 1st.

Eagle's Nest Fairview
TWENTY-TWO PER CENT PROFIT IN TWO WEEKS

Promoted by us last month at 35 cents per share. Now listed on the San Francisco and Salt Lake Exchanges and selling around 45. Good for $1 per share before New Years.

Indian Camp Manhattan
ONE HUNDRED AND FIFTY PER CENT PROFIT IN SIX MONTHS

Promoted by us in April at 30 cents per share. Now selling on San Francisco Stock Exchange at above 75.

Stray Dog Manhattan
EIGHTEEN PER CENT PROFIT IN SEVEN MONTHS

Promoted by us in March at 55 cents per share. Now selling on San Francisco Stock Exchange at 65.

Jumping Jack Manhattan
EIGHTY-TWO PER CENT PROFIT IN EIGHT MONTHS

Promoted by us in February at 30 cents per share. Now selling on San Francisco Stock Exchange at 55.

The five companies are each capitalized for a million shares of $1 each. The entire capitalization is five millions. They were publicly promoted on a valuation of $1,900,000—figured on a basis of the subscription price the public paid for the shares. The listed market value of these standard securities is now $2,900,000, showing an average profit for investors of 52 per cent on their original investment.

We invite correspondence from investors, bankers and brokers.

L. M. Sullivan Trust Co.

Goldfield, - Nevada

Write for our Nevada Mining Securities Review. Sent free on request.

When writing to advertisers please mention "Gossip."

Promoter George Graham Rice was the brains behind the brokerage firm of L. M. Sullivan Trust Company good, $5 will be a reasonable price for them, and if they don't make good, one cent will be too high for them. So why question the ethics of charging 25 cents per share for Seyler-Humphrey when we might have sold it for 15 cents and still have made money; or of charging 15 cents for Manhattan Buffalo when we could have sold it at a profit for 10 cents? The public knows it is gambling. If people want to buy stocks where they won't lose all of their investment under any circumstances, they know they can buy Union Pacific, Pennsylvania Railroad or New York Central. The profits there, however, are limited, just like the losses. In the case of mining stocks, representing prospects under actual development, the public can lose or gain tremendously."

Mr. Elliott, who confessed to me that he often played the horse races when in San Francisco, then wrote out a list of stocks and prices, representing what he said was a book on stocks, comparable to a gambler's book on the the horse races, reading substantially as follows:

Stock	Price	Odds
Union Pacific	$165.00	6 to 5
Reading	155.00	8 to 5
Missouri Pacific	56.00	2 to 1
Erie	28.00	3 to 1
Seyler-Humphrey	.25	20 to 1
Manhattan Buffalo	.15	30 to 1
Manhattan Combination	.10	50 to 1

"There," said Mr. Elliott, "you have the different prices on railroad and mining securities with their chances of winning for the speculator marked against them. When a man goes to a horse race and plays the favorite, he does exactly what the man does who gives his broker an order to buy Union Pacific for him at current quotations. It is about 6 to 5 against the investment making a profit over current quotations on any given day, although the investor will hardly gain 6 for his 5 if the stock enjoys its highest probable advance. It is about 20 to 1 against the man buying Seyler-Humphrey making money, but he will gain 20 for his one if the mine proves to be a bonanza. However, the rail is an investment and the mining a speculation."

"Do you mean to say that the odds against a man making money on Union Pacific on any given day are only 6 to 5 when he buys the stock *on margin?*"

"Not on your life!" he said. "A margin trader on the New York Stock Exchange, unless he has sufficient capital behind him to hold out against inside manipulation, which has for its purpose the shaking out of the speculator, has not got *any* chances! He is bound to lose his money in the end. I am talking about people who buy stocks, pay for them in full and get possession of their certificates and sit tight with them."

Mr. Elliott was a plunger and lost large sums in the gambling houses of Goldfield and Tonopah. He lost $20,000 in a night's play in the Tonopah Club, then owned by George Wingfield and associates. When asked to settle he tendered a check for $5,000 and a certificate for 100,000 shares of Goldfield Laguna Mining Co. stock, then selling at 15 cents. This was accepted. Within a year Laguna sold freely at $2 a share.

This incident illustrates how the foundations were laid for some of the big fortunes which were amassed in the Goldfield mining boom. When George Wingfield came to Tonopah in 1901 he brought with him $150, borrowed from George S. Nixon, then president of a national bank at Winnemucca, Nevada, and later United States Senator. Mr. Wingfield's fortune is now conservatively estimated at between $5,000,000 and $6,000,000.

Success having been won by the Patrick, Elliott & Camp promotions, I was considering whether or not much of the money making that was being done by the promoters around Goldfield was not due to my own peculiar ability to reach the public, and I even meditated on my fitness to become a promoter on my own account. The best properties in Manhattan, by common consent, were the Stray Dog, the Jumping Jack and the Dexter. These were sure enough producers of the yellow metal. They were shippers and were held in high esteem by mining men. I found it impossible to purchase the Dexter because the company was already promoted and the stock widely distributed at around $1 a share. George Wingfield was then and is still interested in the Dexter. The Jumping Jack was unincorporated. The stock of the Stray Dog was practically intact in the hands of the owners. The price asked for the Jumping Jack was $85,000. Stray Dog was held at $500,000.

JUMPING JACK MANHATTAN

I WAS AGAIN IN FUNDS as the result of my profits in the Manhattan boom, and it was again my wont, for want of any other pastime, to play faro at night. I found myself gossiped about with men like January Jones, Zeb Kendall, C. H. Elliott, Al Myers and others who rolled in money one day and were broke the next.

The second largest gambling house in Goldfield was owned by "Larry" Sullivan and Peter Grant, both from Portland, Oregon. Sullivan claimed that he was attracted to Goldfield by the stories which appeared in the Sunday magazine section of a Coast newspaper, the copy for which had been carefully and methodically written in the back room of our Goldfield news bureau. I patronized the Sullivan house, of occasion, and Sullivan usually presided over the games when I was there. One evening I cashed in $2,500 of winnings. The money was piled on the table in $20 gold pieces by the dealer. As I was about to place it in a sack to store away in the safe of the house until the morrow, Sullivan began to josh me like this:

"Say, young feller, why don't you cut me in on some of your mining deals? I'm game!"

"Are you? Well, stack up $2,500 against that money, and I'll see if you are."

He went to the safe and lugged to the table a big canvas sack containing $20 gold pieces. Stacking the money on the table in piles of $400 each, he matched my stake.

"Well?" said he.

"Put that money in a sack," said I, "and go and get that big coonskin coat of yours, take a night ride by automobile to Tonopah, and in the morning go by stage to Manhattan. When you get there look up the owner of the Jumping Jack mine. I have met him. He is a member of the Ancient Order of Hibernians. An Irishman can buy that property from him much cheaper than anybody else. You go and buy it."

"What will I pay?" asked Larry.

"He wants $85,000, but get it as cheap as you can," I replied.

"What? With this $5,000?"

"Yes," said I. "Pay him the $5,000 down and sign a contract to pay the balance in 60 or 90 days; but fetch him back to Goldfield, and have him bring the deeds."

A few days later Sullivan returned to Goldfield, aglow with excitement. Climbing out of the stage coach, he pulled me into his private office.

"Say," he said, "I've got that guy with me and he's got the deeds. I bought the Jumping Jack for $45,000. He'll do anything you want him to do."

"Good!" I said.

The owner was introduced to me, and I turned him over to my lawyer, the late Senator Pyne. Mr. Pyne drew up a paper by which the transferred title of the property to the Jumping Jack Manhattan Mining Co., capitalized for 1,000,000 shares, 300,000 shares of which were placed in the treasury for mining purposes, and 700,000, representing ownership stock, put in escrow, to be delivered to Sullivan and myself on the payment of 6½ cents a share. A board of directors was selected.

At this juncture Sullivan, who knew as much about the mining promotion and mining brokerage business as an ostrich knows about ocean tides, inquired what my next move would be. Sullivan seemed to be bewildered, yet full of faith. My situation was this: I had conceived a rip-snorting promotion campaign for the best property that had yet been offered the public from Manhattan, but I had no cash to present it. Turning to Sullivan I said:

"Do you know the Goldfield manager of the Western Union Telegraph Company?"

"Yes, I know him well."

"Call him up by 'phone or send word to him that you will guarantee payment of any telegrams I file here tonight or during the next three days; I want to send some wires," said I.

"I'll do it," said Sullivan, and within a few minutes I was advised that Sullivan's credit was unquestioned.

I returned to the news bureau and there drafted a 300-word telegram, setting forth the merits of the Jumping Jack Manhattan property and offering short-time options on big blocks of the stock. The message was sent to practically all of the well known brokerage houses in the country which handled mining stocks. The bill for telegraph tolls was $1,200. When Sullivan learned of its size he nearly collapsed.

"How far do you intend to go?" he gasped.

"Well," said I, "how can you lose?" Your friend, Frank Golden, president of the Nye & Ormsby County Bank, has accepted the presidency of the company at our request, and the other officers we have secured are all representative citizens of this community, and, besides, the Nye & Ormsby County Bank has agreed to receive subscriptions. Can you beat that for a layout? Never in my experience in this camp, with all the promotions I have advertised, has the public had a dish quite so palatable offered to it—a producing mine, in the first place; a high class directorate headed by a bank president, in the second place; and a real bank as selling agent, in the third and last place. And it will go like wildfire!"

I labored all that night in my advertising agency on some strongly worded advertising copy recommending to the public the purchase of stock in Jumping Jack Manhattan. In the morning I induced Sullivan to advance $10,000 to pay the advertising bills. The copy was dispatched by first mail to the important daily newspapers of the country, with instructions to publish on the day following receipt of copy.

Within six days all of the advertising had appeared. The effect was magical. The display advertisements assisted the brokers in the various cities, who had asked for reservations of the stock, to dispose of their allotments in a few days. Within ten days after the initial offering of the promotion by telegraph to the Eastern brokers, Sullivan showed me telegraphic orders for 1,280,000 shares of Jumping Jack Manhattan stock at 25 cents a share, an oversubscription of 280,000 shares. Before the stock certificate books were printed and delivered from the local printing office, we were, in fact, oversold.

That week and the next, Sullivan gave me *carte blanche* to speculate in local mining stocks with partnership money, and within a fortnight we had made another small fortune from Manhattan securities. These were advancing in price on the San Francisco Stock Exchange by leaps and bounds.

I recall one overnight winning that we made, amounting to about $12,000, which came so easy I felt almost ashamed to take the money. Manhattan Seyler-Humphrey stock, promoted by Patrick, Elliott & Camp at 25 cents per share, was now listed on the Goldfield and San Francisco stock exchanges. It was in fair demand at 30 cents.

A dispatch reached Goldfield from New York, purporting to be signed by John W. Gates, reading as follows:

> "At what price will you give me an option good 48 hours on 200,000 shares of Manhattan Seyler-Humphrey? Answer to Hotel Williard, Washington, tonight."

This was to Patrick, Elliott & Camp. Within half an hour a half dozen similar messages reached other Goldfield brokers.

I happened to be in the office of Patrick, Elliott & Camp when the first telegram was received, and I lost no time in going out on the street and annexing all the Goldfield offerings of the stock at current quotations. At first Lou Bleakmore, manager for Patrick, Elliott & Camp, smelled a rat, but when he learned that I was buying the stock he became convinced that I believed John W. Gates really wanted some Seyler-Humphrey, and he shot buying orders for his own firm into San Francisco.

Personally I considered the message a snare. Somebody in the East, I guessed, had bitten off a block of Seyler-Humphrey at around 25 cents when it was promoted a few weeks prior and had made up his mind that he would turn a trick. The Goldfield brokers having received telegrams, I assumed that the same message had been sent to brokers in San Francisco, where the stock was also listed. It seemed to me than an advance would certainly be recorded on the following day. Sure enough, the next morning the stock advanced to 38 cents a share, and the market boiled. At this figure, and a little higher, I unloaded in the neighborhood of 100,000 shares in Goldfield and San Francisco. A good deal of this stock had been picked up by me the night before. But I recall that one block of 10,000 shares had been allotted to me weeks before at the brokers' price of 20 cents, and another block of 10,000 shares had been given me as a bonus for my publicity measures.

After turning over to the treasury of the Jumping Jack Manhattan Mining Co. the amount netted from the sale of treasury stock, and paying off the amount still due on the original purchase price, Sullivan and I, within three weeks of my little dare, had cleaned up a net profit of $250,000.

"Do you want a cut?" I asked Sullivan when our joint profits reached the quarter million mark.

"No, I'm game. Stay with it," he returned.

Next day the L. M. Sullivan Trust Co., destined to make and lose millions in the great Goldfield boom that followed and to mold for me an exciting career as a promoter, was formed with a paid-up capital of $250,000. Sullivan was made president and I vice-president and general manager.

ACT TO-DAY

YOU SHOULD KNOW

That mining—legitimate men-and-machinery-mining—the kind carried on by the Calumet & Nevada Consolidated Mines Company in Nevada—yields a higher percentage of profits on capital employed than is shown by **any other** industry.

YOU SHOULD KNOW

That the Calumet & Nevada Consolidated Mines Company is operated in one of the most productive sections of Nevada, the output of this section being figured in the millions.

YOU SHOULD KNOW

That the Company's work in this section has demonstrated that its properties are among the most promising in this part of Nevada.

YOU SHOULD KNOW

The new railroad to Lida greatly enhances the value of the Company's properties and stock.

YOU SHOULD KNOW

That with a railroad at the properties the Company can ship ore running as low as $10 or $15 per ton.

YOU SHOULD KNOW

That the opportunity of investing in this stock is rapidly passing.

ACT TO-DAY

Calumet and Nevada Consolidated Mines Company

Incorporated January 29th, 1908, under the laws of the State of Nevada, and owning and operating the JESSIE TUNNEL, CALUMET & NEVADA and WISCONSIN EXTENSION Properties, Esmeralda County, Nevada.

Capitalization - $2,000,000.00
Divided into 2,000,000 Shares of the Par Value of $1.00 Each

Fully Paid
Non-Assessable
No Preferred Stock

OFFICERS
CASSELL SEVERANCE, President
CHAS. M. LEWIS, Vice-President
E. H. LEWIS, Secretary and Treasurer

DIRECTORS
GEO. L. FAVORITE, Takoma Park, D. C.
AUGUSTUS B. OMWAKE, Hanover, Pa.
CASSELL SEVERANCE, Los Angeles, Cal.
CHAS. M. LEWIS, Los Angeles, Cal.
E. H. LEWIS, Los Angeles, Cal.

REPRESENTATIVES
AUGUSTUS B. OMWAKE, Hanover, Pa.
CHAS. L. EVANS, Williamstown, N. J.
CHAS. M. HEATON, Takoma Park, D. C.
GEORGE W. PARKINS, Cheyenne, Wyo.
W. H. LEWIS, Special Agent.

WHAT TO BUY AND WHAT NOT TO BUY

ONE BEST BUY
EAGLE'S NEST OF FAIRVIEW

BEST BUYS FOR SPECULATIVE-INVESTMENT
Eagle's Nest Fairview
Goldfield Consolidated and Goldfield Kewanas

BEST BUYS FOR SPECULATION
Montgomery Mountain
Tonopah Rescue and North Star

BEST BUYS FOR INVESTMENT
Nevada Hills
Florence Goldfield

THE LATEST "DOPE" ON STANDARD NEVADA MINING SHARES
CLASSIFIED FOR READY REFERENCE BY THE NEVADA MINING NEWS

KEY TO CLASSIFICATION:
1—Best Buys at Present Prices
2—Good Buys at Present Prices
3—Fair Buys at Present Prices
4—May Advance Later
5—Will be Slow to Rise, if at all
6—Don't Buy; Sell if You Own Any

TONOPAHS

Classification	STOCKS	High 1907	Feb. 6, 1908 P. M. Quotations Bid	Ask	REMARKS
1	NORTH STAR	$3.90	$.18	$.19	Strike verified. Has extension of Montana ledge. Buy it for quick profits. Will surely sell higher soon.
1	RESCUE	.25	.15	.16	Experts declare this one will open up the Belmont vein. Belmont property adjoins. Work going on at 600 level on Rescue. Buy it for quick profits.
2	BELMONT	5.50	1.72½	1.75	Big strike reported in this property. Good buy on reactions.
2	MONTANA	3.90	1.92½	1.95	New strike reported. Intrinsically worth double present price. Buy it for sure profits.
2	TONOPAH OF NEVADA	20.00	5.30	5.50	One of the greatest mines in the West. Buy it and hold. Is worth $10 per share.
3	MIDWAY	2.10	.80	.84	Intrinsically one of the cheapest of the Tonopahs. Entitled to a big rise.
4	JIM BUTLER	1.30	.44	.45	Not making any money.
4	EXTENSION	5.50	...	1.00	Probably worth more, but market is narrow.

GOLDFIELDS

Classification	STOCKS	High 1907	Feb. 6, 1908 P. M. Quotations Bid	Ask	REMARKS
1	GOLDFIELD CONSOLIDATED	10.50	5.00	5.02	Good for $10 now that the weak holders have been shaken out by Baruch and his crowd of his bidders. Don't buy it on margin. Buy it outright and hold it.
1	KEWANAS	1.75	.51	.53	Has reacted with the leaders, but is the strongest them all.
1	FLORENCE	5.02½	4.45	4.50	Big advance in price recently does not yet remove one from bargain counter. At present prices, compared with others and considering mine showing, equipment, management and treasury reserve, the cheapest of the higher priced Goldfields.
2	FLORENCE EXTENSION	.65	.32	.34	A good prospect. Is being manipulated for higher. A shipper.
2	ST. IVES	1.10	.45	.50	Has big acreage and is a fine prospect.
2	SILVER PICK	1.30	.31	.33	Same as Silver Pick, but more subject to manipulative influence.
2	RED HILLS	.80	.32	.34	Being pushed up. Not a bargain.
2	DAISY	6.15½	1.02½	1.05	A good prospect with fine acreage. Worth more.
2	PORTLAND		.13	.14	

3. A Saturnalia of Speculation

Mr. sullivan's gambling house affiliation was not considered a drawback to the trust company. George Wingfield, vice-president and heaviest stockholder of the leading bank in Goldfield, was a gambler and Mr. Wingfield also owned extensive interests in the mines. His mines were making good, too. Owners of the gambling places now stood as much for financial solidity in Goldfield as did savings bank directors in the East.

As for myself, I was unafraid. I vowed I would henceforth prove an exception to the mining camp rule and quit all forms of gambling. My new position demanded this. And I found it easy to obey the self-imposed inhibition. Soon the stock market operations of the trust company gave my speculative instinct all the vent it could possibly have craved under any circumstances.

A few days later the sobering sense which impelled me to resolve that I must absent myself from gambling tables evolved into a stern ambition to accomplish big things for the trust company. I went about my business like a man who sees dazzling before him a golden scepter and who is imbued with the idea that if he exerts the power he can grasp the prize. It had been agreed that the trust company would specialize in the promotion of mining companies, and I determined that the trust company should conduct its business as a trust company ought.

John Douglas Campbell, known on the desert as plain "Jack" Campbell, was engaged by the trust company as its mining adviser and mine manager. We agreed to pay him a salary of $20,000 a year, with a bonus of stock in every new mining company we promoted, a stipend which was later found to be equivalent to $50,000 a year.

Mr. Campbell had been identified with Tonopah and Goldfield mining interests for three years, and was favorably known. For eight years before coming to Tonopah he was employed as a mining superintendent in Colorado by Sam Newhouse, the multi-millionaire mine operator of Utah. In Colorado Mr. Campbell's reputation had been good. On coming to Tonopah he was employed by John McKane, then associated with Charles M. Schwab. Later he was placed in charge of the Kernick and Fuller-McDonald leases on the Jumbo mine of Goldfield from which, during a year's time, $1,000,000 in gold was taken out. After that Mr. Campbell took hold of the Quartzite lease at Diamondfield, near Goldfield, and he produced $200,000 in a few months from that holding. He followed this up by a record production from the famous Reilly lease on the Florence mine of Goldfield, amounting to $650,000 in two months. It was within thirty days of the date of expiry of the Reilly lease that Mr. Campbell was induced to take charge of the mining department of the trust company.

Mr. Campbell's advent as our mine manager was immediately reflected in the stock market by the advance of Jumping Jack Manhattan Mining Co. shares, which were now regularly listed on the San Francisco Stock & Exchange Board, to 40 cents per share, up 15 points from the promotion price. The sharp rise wrought an undoubted sensation in stock market circles. Brokers in the cities who had sold Jumping Jack to their customers clamored for a new Sullivan promotion. Any new mining venture for which the trust company would stand sponsor was assured of heavy subscription and a broad public market.

TRYING IT ON THE STRAY DOG

THE STRAY DOG MANHATTAN mine was furnishing daily sensations in the way of frequent strikes of fabulously rich ore. I urged that, no matter how small the profit, the Sullivan Trust Co. should begin its corporate career with the promotion of a property as good as the Stray Dog. The Stray Dog was for sale—at a price. One interest, of 350,000 shares, owned by Vermilyea, Edmonds & Stanley, the law firm of highest standing in Goldfield, could be acquired at 45 cents a share, and another interest, of 350,000 shares, owned by prospectors who had located the ground, could be had at 20 cents a share, all or none. The remainder of the stock was in the treasury of the company. The total demanded for 700,000 shares of ownership stock was $227,500, all cash. A likely property adjoining the Stray Dog, known as the Indian Camp, could be purchased for $50,000 in its entirety. We knew that as soon as it should become known that we had bought the Stray Dog, the value of Indian Camp ground would double, and we therefore decided to annex the Indian Camp at the same time we took over the Stray Dog.

The proposed outlay amounted to more money than we had, and I looked about for assistance. Henry Peery, a Salt Lake mining man of substance, had been negotiating for the Stray Dog in the interest of Utah bankers. We agreed that Mr. Peery should be allowed to participate on the basis of a one-third interest for him, and a two-thirds interest for the trust company. Besides supplying his quota of the cash needed to swing the deal, Mr. Peery agreed to furnish a president for the company, who, he said, interested himself very frequently in mining enterprises. This was Henry McCornick, the Salt Lake banker, son of the head of the firm of McCornick & Co., reputed to be the richest private bankers west of the Mississippi River. The deal was made.

We immediately proceeded to promote the Stray Dog Manhattan Mining Co. at 45 cents per share, the average cost to us of the stock being $32\frac{1}{2}$ cents. It was impossible for any huge profit to accrue in Stray Dog on any such margin as $12\frac{1}{2}$ cents per share between our cost price and the selling price, because the expense of promotion appeared bound almost to equal this. We figured that any promotion profits must come out of the Indian Camp. The Indian Camp was capitalized for 1,000,000 shares, 650,000 of which were paid over to the trust company and to Mr. Peery for the property. The remaining 350,000 shares were placed in the treasury of the company to be sold for purposes of mine development. The average per share cost to the trust company of its ownership stock was a fraction less than eight cents. We decided that as soon as the Stray Dog was promoted we would offer Indian Camp shares on a basis of 20 cents per share net to the brokers and 25 cents to the public, and looked forward, if successful, to gaining about $75,000 net on both ventures.

Immediately on taking over the control of Stray Dog and Indian Camp the trust company purchased treasury stock in each of these companies, and put a large force of men to work to open up the properties. Within thirty days of the incorporation of the trust company Gold Hill in Manhattan, on which were located the Stray Dog, Jumping Jack and Indian Camp, swarmed with miners. The orders given to Engineer "Jack" Campbell were to put a man to work wherever he could employ one, and to be unsparing in expense so long as he could obtain results. Towering gallow frames and 25-horsepower gasoline engines were installed and other necessary mining equipment ordered shipped to the properties. Blacksmith shops, bunk houses and storehouses were erected on the ground. Day and night shifts of miners were employed. In order to guarantee the constant presence on the properties of the engineer in charge, the Sullivan Trust Co. built for the engineer's use a $6,000 dwelling house on Indian Camp ground.

Having convinced the natives that we were in dead earnest about our

Jumping Jack Manhattan Mining Co.

In February, 1906, the management of the company launched its first promotion. It was the Jumping Jack Manhattan, and was placed on the market at 30 cents a share. This was promptly taken up and the stock over-subscribed for. It was very shortly listed on the San Francisco Exchange and the New York Mining Exchange, and within one week from the close of the subscription books Jumping Jack Manhattan was eagerly sought at 40 cents a share. Since then it has advanced to 54 cents, at which price it is quoted as we go to press. The properties owned by the Jumping Jack Manhattan Mining Company are the Jumping Jack claims in Manhattan, adjoining the Stray Dog and Union No. 9. All three of these properties are shippers of high-grade ore. At the time the Jumping Jack Manhattan Mining Company purchased the Jumping Jack claim, the mine was shipping ore to the smelters and regular returns were being made to the treasury of the company. Since then the company has continued its development work and shipping, and the showing thus far made substantiates the favorable report made by the company's engineering corps. A vein as wide as a room in an ordinary dwelling-house has been opened up at the 150-foot level, and returns gratifying assay values in milling ores. The equipment of the mine consists of a 35-h. p. gasoline hoist and the necessary houses and buildings for the conduct of the company's business. Sinking on the shaft has been vigorously prosecuted, and it is the intention of the management to continue to the 300-foot level, which they hope to reach by the first of the year. From this point crosscuts will be started to again cut the vein and ascertain its width at that depth.

Stray Dog Manhattan Mining Co.

The success of the L. M. Sullivan Trust Company's first promotion in the district was so marked that a second property was secured immediately, adjoining the Jumping Jack. This was incorporated by Vermilyea, Edmonds & Stanley as the Stray Dog Manhattan Mining Company, and to secure control of this valuable property the L. M. Sullivan Trust Company paid Vermilyea, Edmonds & Stanley $160,000 cash for 320,000 shares of the stock. This stock, added to the stock of the original locators of the claim, which the L. M. Sullivan Trust Company secured at the time they bought the Jumping Jack, gave absolute control. Stray Dog stock was floated at 55 cents and has since sold up to 74 cents a share, that being the quotation last Thursday. The highest grade ore yet found in Manhattan has been opened up on the Stray Dog. At the 135 foot level, a crosscut disclosed an 8-foot vein of ore which gives the handsome average of about $57 to the ton, with streaks running through this 8-foot vein which will run as high as $1000 a ton, or 50 cents a pound. Work on the Stray Dog is being pushed with the energy characteristic of any properties coming under the influence of the progressive management of the L. M. Sullivan Trust Company, and every day brings forth interesting developments as the property is further exploited and opened up.

L. M. Sullivan Trust Co. properties

Advertising Department

In one man's hand is placed the entire department of advertising, and some idea of the duties devolving upon him may be gained from knowledge of the fact that he spends more than $60,000 a month for newspaper publicity. Many of the company's announcements are of such importance that whole pages are wired to big New York, Chicago and San Francisco newspapers, and the company's Western Union bill alone frequently amounts to more than $10,000 in a single week. The office force includes six expert book keepers who are paid fancy salaries, and are the best obtainable, while a corps of sixteen stenographers drawing $50 a week each attend to the correspondence of the company, which is tremendous in volume.

mine making intentions, we busied ourselves offering Stray Dog stock for subscription at 45 cents per share. It was well known around the camp that we had paid 45 cents per share for one block of 350,000 shares, and mining camp followers were among the first to subscribe for the stock. Then an effort was made to dispose of quantities of it to the Eastern public by advertising and through mining stock brokers.

That advertising campaign was approached with considerable caution. In the first place, the subscription price of Stray Dog, 45 cents, was 80 per cent higher than that of any other advertised promotion which had yet been made from either the Goldfield or Manhattan camps; and in the second place, the conduct of a mining stock promotion campaign by a trust company appeared to me to justify more than ordinary care. There were other factors that entered for the first time in Goldfield, too.

The initial successes of the big display advertising campaigns directed from Goldfield appeared to have been due to the fact that the American public had greeted mining stock speculation as filling a long felt want, namely, a channel for speculation in which they could indulge their gambling spirit with limited resources—resources that were insufficient to give them a look-in on the big exchanges where the high priced rails and industrials are traded in.

ADVERTISING FOR THINKERS

HAVING "TRIED ON THE DOG" my methods of advertising for nearly two years, that is to say, having conducted an advertising agency for mine promoters, and learned the business with their money, I had passed through the experimental stage and now marshalled a cardinal principal or two that I decided must guide me in the operations in which I had become more directly interested.

I resolved never to allow an advertisement to go out of the office that was unconvincing to a thinker. If my argument convinces the man of affairs, I determined, it will certainly win over the man of no affairs.

Dogmatically expressed, the idea was this: Never appeal to the intelligence of fools, no matter how easily they may part with their money. Turn your batteries on the thinking ones and convince them, and the unthinking will to follow. That principle was applied to the *argument* of the advertisement.

The headlines were constructed on an entirely different principle, namely, to be positive to an extreme. The Bible was my exemplar. It says, "It is" or "It was," "Thou shalt" or "Thou shalt not," and the Bible rarely explains or tells why. The strength of a headline lies in its positiveness.

The logic which directed that the flaring headline of my big display advertising copy embrace a very positive statement, and that the *argument* which followed in small type be convincing to the thinker, was based on a recognition of the fact, that, while boldness of statement invariably attracts attention, analysis is the final resort of the thinker before becoming convinced.

More circumspection was used also in the process of selecting media for the advertising. Newspapers that did not publish in their news columns mining stock quotations of issues traded in on the New York Curb, the Boston Stock Exchange, the Boston Curb, the Salt Lake Stock Exchange or the San Francisco Stock Exchange were taboo, on the theory that by this time trading in mining stock had grown sufficiently popular to command a regular following, and that it was easier to appeal to those who had some experience in mining stock speculations than to those who had never before ventured.

Subsequent advertising campaigns were always conducted from this viewpoint. I did not set the ocean on fire with my Stray Dog promotion, the advertising campaign of which was conducted on these lines, but this was due to circumstances which I explain further on. Later, when the Sullivan Trust Co. grew and prospered, and afterward when I reached the East and learned more and more of the inside mechanism of the big Wall Street promotion game in rails and industrials as well as mining stocks, I found that my publicity principles were comparable to those accepted by the Street generally.

The mighty powers of Wall Street recognize the fact that it is not in the nature of things that fools should have much money, and thinkers, not fools, are the quarry of the successful modern day promoter, high or low, honest or dishonest.

A little knowledge is a dangerous thing, and the man who thinks he knows it all because he has accumulated much money in his pet business enterprise is a typical personage on whom the successful modern day multimillionaire Wall Street financier trains his batteries.

The honest promoter aims at both the thinker who thinks he knows but doesn't, and the thinker who really does know. He is compelled to appeal to both classes because the membership of the first outnumbers that of the second in the propotion of about 1,000 to 1.

In fine, for every dollar of "wise" money which is thrown into the vortex of speculation, $1,000 is "unwise," or considered so.

The initial Stray Dog and Indian Camp promotion campaign was only half successful at the outset. About 650,000 shares of Stray Dog and 350,000 shares of Indian Camp had been disposed of when the Manhattan boom began to lose its intensity. Promotions had been made a little too rapidly for public digestion. There were more miners at work than ever in the Manhattan camp, but the demand for securities was not keeping pace with the supply. Manhattan's initial boom appeared to be flattening out just as Goldfield's first boom had.

We met with a setback from another direction. Henry McCornick's banking connections in Salt Lake objected to the use of his name as president of the Stray Dog. At the very height of our advertising campaign Mr. McCornick resigned. We elected our engineer "Jack" Campbell, president, but damage was done.

YES, "BUSINESS IS BUSINESS"

THE OFFICES OF THE trust company were furnished on an elaborate scale, resembling the interior of a banking institution of a large city. The offices became the headquarters of Eastern mining stock brokers whenever they arrived in camp.

One morning J. C. Weir, a New York mining stock broker, whose firm held an option from the trust company on 100,000 shares of Stray Dog stock, was ensconced in one of the two luxuriously furnished rooms used as executive offices. Mr. Weir's firm was one of our selling agents in New York. He was the dean of mining stock brokers in New York City. In those early days the telephone service of Goldfield was not yet perfected, and it was only necessary for a person, in order to overhear any talk over the telephone in our offices, to lift the receiver from the nearest hook and listen. It was reported to me that Mr. Weir had been availing himself of this method of learning things at first hand.

"Say, Rice," said Mr. Sullivan one morning, "Weir hears your messages every time you are called on the 'phone. He takes advantage of you. I wish you would let me fix him."

"All right; what do you want to do?" I answered.

"Say," said Mr. Sullivan, "Campbell, our engineer, is in Manhattan. I'll call him up from the public station and tell him to 'phone you some red hot news about mine developments on Stray Dog, and I'll see to it that Weir is in his office at the time you get the message. If Weir don't steal the news and grab a big block of Stray Dog on the strength of it, I'm a poor guesser."

All of our options to brokers were to expire on the 15th of March and this was the 13th.

At four o'clock in the afternoon I was in my room. Mr. Weir was at the desk in the room opposite. The 'phone bell rang.

"Hello," I said, "who is this?"

"Campbell, at Manhattan," was the response.

"What's the news, Jack?" I asked.

"We've just struck six feet of $2,000 ore! It's a whale! Never saw a mine as big as this one in my life! Don't sell any more Stray Dog under $5 a share!" shouted Mr. Campbell.

"Bully, Jack," I said, "but keep that information to yourself. Don't tell your mother, and don't let any more miners go down the shaft. Close it up until I am able to buy back some of the stock I sold so cheap."

Fifteen minutes later Mr. Sullivan and I met Mr. Weir leaving the room.

"Weir," said I, "your option on Stray Dog expires on the 15th at noon. So far, your New York office has ordered only 85,000 shares of the 100,000 that were allotted to you. We have decided to close subscriptions on the moment and wish you would wire your New York office not to sell any more."

"You are wrong," said Mr. Weir; "why, when I left New York we had oversold our entire allotment! If the office has not notified you of this, it has been a slip. We will, in fact, need at least 25,000 shares more."

"You can't have them," said I.

"Not in a thousand years!" put in Mr. Sullivan.

Mr. Weir sent a bunch of code messages to New York. All the next day Mr. Sullivan spent with Mr. Weir. He allowed Mr. Weir to cajole him into letting him have the entire block of stock. Finally, it was agreed beween Mr. Weir and Mr. Sullivan that Mr. Sullivan would give him the additional stock whether I consented or not. Surreptitiously, according to Mr. Weir's idea, Mr. Sullivan was yielding to him, without my knowledge and against my wishes.

Next day the Sullivan Trust Co. shipped to Mr. Weir's firm in New York 25,000 shares of Stray Dog attached to draft at 45 cents a share. The draft was paid. The avenging angel kept hot on Mr. Weir's trail, for right on the heels of the New York broker's Stray Dog purchase came a calamity which almost obliterated the market values of Nevada mining stocks and particularly those of the shares of Manhattan mining companies. San Francisco was destroyed by earthquake and fire. Not less than half of the capital invested in Manhattan stocks had come out of the city of San Francisco. The earthquake was fatal to Manhattan.

The San Francisco Stock Exchange, which was the principal market for Manhattan mining shares, was compelled to discontinue business for more than two months. Brokers and transfer companies lost their records, and the Coast's property and money loss was so appalling that no more money was forthcoming from that direction for mining enterprises. Every bank in Nevada closed down, just as every California bank did, the Governors of both states declaring a series of legal holidays to enable the financial institutions to gain time. Nevada banks, as a rule, had cleared through San Francisco banks, and practically all of Nevada's cash was tied up by the catastrophe.

The Sullivan Trust Co. faced a crisis. I had decided it was good business to lend support to Jumping Jack in the stock market when the Manhattan boom began to relax from its first tension, and had accumulated several hundred thousand shares at an average of 35 cents. The trust company had only $8,000 in gold in its vaults on the day of the 'quake. Moneys deposited in bank were not available. Of the $8,000 in gold coin, $6,500 was paid two days after the earthquake to the Wells-Fargo Express Co. for an automobile which was in transit at the time, and for which Wells-Fargo demanded the coin. It was impossible to hypothecate mining securities of any description in Nevada or San Francisco. With the Sullivan Trust Co.'s funds tied up in closed-up banks, and with an unsalable line of securities in its vaults, it was "up against it."

For a period it looked as if we must go to the wall. For two months we eked out a bare subsistence by the direct sale of Manhattan securities at reduced prices to the Eastern brokers. This purchasing power came largely from brokers who were short of stocks to the public on commitments made at a much higher range of prices and needed the actual certificates for deliveries.

It took the Nevada banks and the San Francisco Stock Exchange more than sixty days to rehabilitate themselves. No sooner did the San Francisco Stock Exchanges open for business than it became possible for the Sullivan Trust Company to borrow some much needed cash on Manhattan securities, of which it had a plethora. Through members of the San Francisco Stock Exchange, it obtained in this way in the neighborhood of $100,000. Goldfield banks supplied another $100,000 a little later by the same process. Then the clouds rolled by.

FORTUNES THAT WERE MISSED

Soon the Mohawk of Goldfield began to give unerring indications of being the wonderful treasure house it has since proved to be. Hayes and Monnette, who owned a lease on a small section of the property, had struck high-grade ore and were producing at the rate of $3,000 per day. A few weeks later it was reported that the out-put had increased to $5,000 a day.

The Mohawk being situated only a stone's throw from the Combination mine, the idea that the Mohawk might turn out to be another Combination

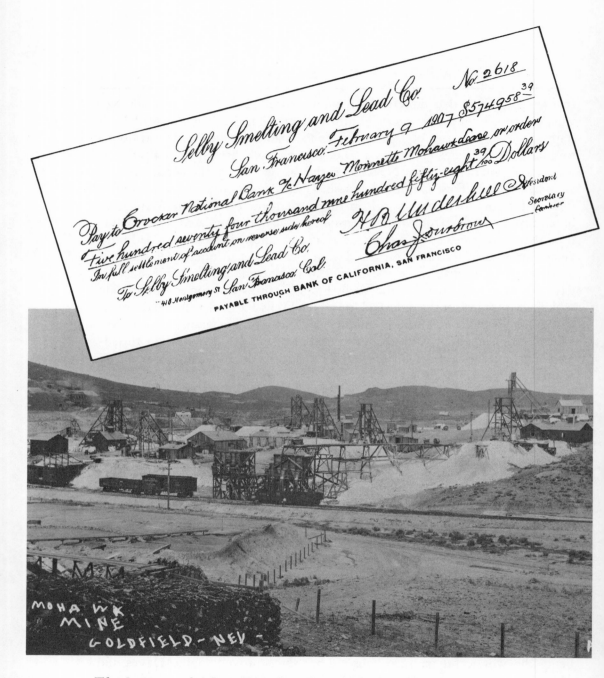

The $574,900 check was issued to the Mohawk Mine in settlement for 47 tons of ore shipped from the Hayes-Monette lease. All was part of the Goldfield Consolidated properties.

was common in Goldfield. Hayes and Monnette were startled—almost frightened—at their success. Yielding for the moment to the warning of friends, who urged upon them the possibility of the ore soon pinching out, Hayes and Monnette called at the offices of the trust company and offered to sell their lease, which had six months to run, for $200,000 cash and $400,000 to be taken out of the net proceeds of the ore. My gambling instinct was aroused.

"I will take it," I said.

I sent over to the State Bank & Trust Co., and had a check certified for the $200,000. I was about to close the deal when Mr. Sullivan and "Jack" Campbell protested.

"I ought to have fifteen days to examine the mine," urged Mr. Campbell.

"It is too big a chance to take," declared Mr. Sullivan.

When appealed to, Hayes and Monnette said that to allow a fifteen day examination would mean practically to shut down the property for that period and would result in a positive loss to them because of the limited period of their lease. The extent of the loss, if the deal fell through, was too large to contemplate, and they refused.

Day by day, as Mr. Campbell and Mr. Sullivan dilly-dallied, the output of the lease increased, and when, a fortnight later, all three of us were unanimously in favor of the proposition, Hayes and Monnette flatly refused to sell. Within half a year that lease on the Mohawk produced in the neighborhood of $6,000,000 worth of ore gross, and netted the leasers about $4,500,000. The Sullivan Trust Co. certainly overlooked a bet there.

During this period I spent an evening with Henry Peery and W. H. "Daddy" Clark. Mr. Clark, like Mr. Peery, hailed from Salt Lake. Mr. Clark had successfully promoted the Bullfrog Gibraltar. Seated around a table in the Palm Restaurant, the conversation turned to new camps.

"Rice," said Mr. Clark, "I expect to be able to put you in on a townsite deal in a couple of weeks that will make you some money if you undertake to give the camp some publicity."

"Good," said I.

"I am having some assays run," he said, "of some samples which were brought into camp last night by a couple of prospectors, and if they turn out to be what the prospectors claim, or anything near it, we'll need your services to put a new camp on the map."

That night Mr. Peery learned from the assayer that the lowest assay of 16 samples was $86, and the highest $475, per ton. Next morning Mr. Peery informed me that he had remained all night with Mr. Clark to learn where the ore came from. Mr. Peery said that Mr. Clark had told him, in the wee sma' hours, that the place was Fairview Peak, fifty miles east of Fallon.

"Rice," said Mr. Peery, "let's beat him to it. He's going to trek it across the desert by mule team with a camp outfit tomorrow, and it will take him a week to get there."

"Billy" Taylor, who was interested with Mr. Peery in a Bullfrog enterprise, joined the party, and we each gave Mr. Peery a check for $500, forming a pool of $1,500 to send a man to Fairview to buy properties there. Mr. Peery wired the Bank of Republic at Salt Lake to pay Ben Luce $1,500, and instructed Mr. Luce by wire to take the money, go to Fairview and do business.

It was nearly two weeks before we heard of either Mr. Clark or Mr. Luce. Mr. Clark returned to camp and said he had purchased from a group of itinerant prospectors the Nevada Hills property, scene of the big find, for $5,000, and that it was a "world beater."

"Did you meet any outsiders there?" queried Mr. Peery.

"Yes," said Mr. Clark, "I met a man named Luce who almost got ahead of me. In fact, he did buy the property before I got there, but he had no money, and they would not take his check for $500, which was the deposit required. I had the gold with me, and that settled it."

A few days afterward, Mr. Luce came to Goldfield.

"I didn't get the big one," he said, "but I bought the Eagle's Nest, nearby, for $7,000, of which $500 was demanded to be paid down, and there is ore in it and it looks good to me. I had no money with me when I arrived in Fairview. They refused my check for the Nevada Hills, but the Eagle's Nest boys took it for their first payment of $500."

Mr. Luce was not at home when Mr. Peery's despatch was delivered in Salt Lake. When it reached him the bank was closed. In order to catch the first train he was compelled to leave the money behind. He arrived in Fairview minus the $1,500, and thereby lost the Nevada Hills for Mr. Peery, Mr. Taylor and the Sullivan Trust Co.

Mr. Clark and his partners incorporated the Nevada Hills for 1,000,000 shares of the par value of $5 each and accepted subscriptions at $1 per share.

Within a few months the Nevada Hills paid $375,000 in dividends out of ore, and soon thereafter, at the height of the Goldfield boom, it was reported that the owners of the control refused an offer of $6,000,000 for the property. The mine has turned out to be a bonanza. The stock of the company sold recently on the New York Curb and San Francisco Stock Exchange at a valuation for the mine of $3,000,000, and it is believed by well-posted mining men to be worth that figure. George Wingfield, president of the Goldfield Consolidated who followed the Sullivan Trust Co. into Fairview and bought the Fairview Eagle, which is sandwiched in between the Nevada Hills and the Eagle's Nest, is now president of the Nevada Hills. Treasury stock of the Fairview Eagle was sold in Goldfield at 40 cents per share. Recently the Nevada Hills and Fairview Eagle companies were merged.

"Jack" Campbell reported favorably on the Eagle's Nest, and we decided to organize and promote a company to own and develop the property.

The Sullivan Trust Co. bought Mr. Taylor's interest in the Eagle's Nest for $8,000, Mr. Luce's for $8,000 (he had been awarded a quarter interest for his work), and Mr. Peery's for $30,000. It made the property the basis for the promotion of the Eagle's Nest Fairview Mining Co., capitalized for 1,000,000 shares of the par value of $5 each. Governor John Sparks accepted our invitation to become president of the company. The entire capitalization was sold to the public through Eastern and Western stock brokers within thirty days at a subscription price of 35 cents per share. After paying for the property, our net profits were in the neighborhood of $150,000.

The Eagle's Nest deal enabled the trust company to repay most of the money it had borrowed after the San Francisco earthquake and put the company on Easy Street again.

THE TALE OF BULLFROG RUSH

FOLLOWING THE EAGLE'S NEST promotion, the Sullivan Trust Co. became sponsor for Bullfrog Rush. I had met Dr. J. Grant Lyman, owner of the property, on the lawn of one of the cottages of the United States Hotel in Saratoga a few years before, where he raced a string of horses and mixed with good people, and I knew of nothing that was to his discredit. Dr. Lyman bought the Bullfrog Rush property for $150,000. I was present when he paid $100,000 of this money in cash at John S. Cook & Co.'s bank in Goldfield. The Bullfrog Rush property was composed of large acreage, enjoyed splendid surface showings, and was situated contiguous to the Tramps Consolidated, which was then selling around $3 a share. It looked like a fine prospect.

Dr. Lyman incorporated the company for 1,000,000 shares of the par value of $1 each. The services of the Sullivan Trust Co. were employed to finance the enterprise for mine development. The Trust company obtained an option on the treasury stock of the company at 35 cents per share, and proceeded to dispose of it through Eastern brokers and direct to the public by advertising, at 45 cents per share to brokers and 50 cents per share to investors. We sold 200,000 shares, realizing $90,000 in less than thirty days, retained $20,000 for commission and expenses, and turned into the treasury of the

Indian Camp Manhattan Mining Co.

The Indian Camp Manhattan Mining Company was next on the list of promotions made by the L. M. Sullivan Trust Company. This property adjoins the Union No. 9 mine of the Dexter Company and also the Little Grey, both of which are proved mines. The Indian Camp has always been considered a low-grade proposition of milling values only, and has proved to be such, as in fact, are most of the Manhattan properties, for Manhattan is not a high-grade camp, like either Tonopah or Goldfield, but its ores are of such tremendous volume that the camp and district are among the best in the State. The Indian Camp property is a typical Manhattan mine, and already immense bodies of ore have been blocked out and in such quantities that Indian Camp stock has experienced the highest advance of any of the Sullivan flotations. It was promoted at 30 cents a share in April, just a few days previous to the San Francisco catastrophe, and despite the fact that the San Francisco disaster had a greater effect on Manhattan stocks than any other because of the tremendous amount of California capital invested there, Indian Camp stock is now selling around $1.05, showing a net profit of 275 per cent in six months to investors.

Bullfrog Rush Mining Co.

Located in the Bullfrog district, with excellent surface and upper tunnel showings, and adjoining some of the best properties of the Bullfrog district, are the properties acquired by the Bullfrog Rush Mining Company, which was floated by the L. M. Sullivan Trust Company, following their last Manhattan promotion. A large body of low-grade ore was disclosed on the higher levels of the Bullfrog Rush property at the time the company acquired the property, but since then development work showed that the ore had pinched and the best values now obtained are 60 and 70 cents a ton. The Sullivan Trust Company, as a result, took immediate steps to notify their clients that their money would be refunded on stock of the Bullfrog Rush Mining Company purchased through the Sullivan Company. This action on the part of the management is unprecedented in the history of the promotion business of the West, and has done more to establish confidence in the company and the State than any other act of a firm or corporation. Had the mine proved up with depth, the stockholders would have earned tremendous profits, but since it proved otherwise, the L. M. Sullivan Trust Company shouldered the burden and refunded the cash, at a loss exceeding more than a hundred thousand dollars.

Eagle's Nest Fairview Mining Co.

Ever alert for new fields and opportunities, the Sullivan Company was early on the ground in the morning of Fairview's fame. This coming great gold camp is located in Churchill County, within fifty miles of the main line of the Southern Pacific Railroad, and was discovered last fall by F. O. Norton, an old-time prospector of Nevada. Practically every claim among the first locations paid from the "grass-roots" and stocks that were placed on the market at a few cents in some of the companies have now reached a market value of $3.50. Admirably located in the very heart of the most richly mineralized section of the Fairview district, lies the claims secured by the L. M. Sullivan Trust Company and incorporated under the title of the Eagle's Nest Fairview Mining Company. This flotation was made at 35 cents a share, and within a week after the subscription books were closed in October, the stock was selling on the San Francisco Stock and Exchange Board at 45 cents, and is now selling around that figure. George Wingfield, one of the heaviest and most successful mining operators in the State, has purchased a large block of stock in the company and par is predicted for it by the first of the year. It is unquestionably one of the four great properties of Fairview, and compared with the prices at which other listed Fairview stocks are selling, it is considered the best buy on the Fairview market.

The Eagle's Nest Fairview Mining Company owns a large acreage on the hills above the town of Fairview, and is a close neighbor to the Nevada Hills, the Dromedary Hump, the Fairview Eagle and the Hailstone properties, all of which are being actively developed and rapidly placed on the list of shippers; in fact, the Nevada Hills has been shipping fifty-eight miles to the railroad all summer. The properties of the Eagle's Nest Fairview are considered among the surest producers on the Sullivan list of promotions, and they are being closely watched by mining men of the State and the West.

L. M. Sullivan Trust Co. properties

Bullfrog Rush company $70,000, all of which was placed at the disposal of the company for mine development.

Half a dozen tunnels were run and several shafts were sunk. Down to the 400-foot level the mine appeared to be of much promise. It was then learned that the shaft at the 400-foot point had encountered a bed of lime. It appeared that all the properties on Bonanza Mountain, where the Bullfrog Rush was situated, including the Tramps Consolidated, which was then selling in the market at a valuation of $3,000,000, were bound to turn out to be rank mining failures. The entire hill, according to our engineer, was a slide, and below the 400-point ore could not possibly exist.

We thereupon notified Dr. Lyman that we would discontinue the sale of the stock until such time as the property gave better indications of making a mine.

A few weeks later Dr. Lyman entered my private office unannounced. At this period Jumping Jack, Stray Dog, Indian Camp, and Eagle's Nest were all selling on the San Francisco Stock Exchange at an average of 35 per cent above promotion prices. The L. M. Sullivan Trust Co. was making good to investors. Bullfrog Rush had not yet been listed, and we were afraid to give it a market quotation.

"I have formed here in Goldfield the Union Securities Co.," Dr. Lyman said, as he sat down close to my desk, "and I am going into the promotion business myself. I don't believe a word of the report you have that the Bullfrog Rush is a failure. I am going on with the promotion."

I protested. "We shall not permit it," I said. "Governor Sparks, who is the best friend the Sullivan Trust Co. has, accepted the presidency of the Bullfrog Rush on our assurance that the property was a good one. John S. Cook, the leading banker of this town, accepted the treasurership on the same representations. Mr. Sullivan, president of this trust company, is vice-president of the Rush. We are 'in bad' enough as the matter already stands. Don't dare go on with the promotion at this time."

Dr. Lyman left the office without uttering a word.

Two days later I received a dispatch from Governor Sparks saying that a full page advertisement of the Union Securities Co. had appeared in the *Nevada State Journal* at Reno, offering Bullfrog Rush stock for subscription. The Governor protested vigorously against the sale of the stock. We had previously informed him as to the new conditions which prevailed at the mine.

I sent Peter Grant, one of Mr. Sullivan's partners in the Palace, to Dr. Lyman to protest. The answer came back that the *Nevada State Journal* advertisement was about to be reproduced in all the newspapers of big circulation throughout the East, and that the orders for the advertisements would not be canceled. Half an hour later Dr. Lyman entered the office with Mr. Grant. Mr. Grant looked nettled. Dr. Lyman glowered.

I bade Dr. Lyman take a chair.

"If you move a finger to stop me," he said, as he sat himself down before me, "I'll expose every act of yours since you were born and show up who the boss of this trust company is!"

Dr. Lyman was tall as a poplar and muscled like a Samson. He was fresh from the East, red-cheeked and groomed like a Chesterfield. I was cadaverous, desert-worn, office-fagged, and undersized by comparison. In a glove fight, Dr. Lyman could probably have finished me in half a round. But the disparity did not occur to me. The sense of injustice made me forget everything except Dr. Lyman's blackmailing threat. I jumped to my feet. Dr. Lyman backed up to the glass door. I aimed a blow at him. He backed away to dodge it. In a second he had collided with the big plate glass pane, which fell with a crash. In another instant he recovered his feet, turned on his heel and ran. His face was covered with scratches, the result of his encounter with the broken plate glass. Several clerks who followed him, thinking he had committed some violent act, reported that he didn't stop running until he reached the end of a street 600 feet away.

"Oh," he gasped, "I never want to see such a look in a man's eyes again. I thought I saw him reach for a gun."

Such an idea was farthest from my mind, although I was very angry. Conscience had made a coward of the doctor. I was quick to decide upon a course of action.

The position of the trust company was this: with the exception of Bullfrog Rush, we had a string of stockmarket winners to our credit with the public. If we allowed Dr. Lyman to go ahead with his promotion of Bullfrog Rush, we should, unless we abandoned our rule to protect our stocks in the market, be compelled some day to buy back all of the stock he sold. The truth about the mine was bound to come out, and we stood before the public as its sponsors.

I decided that the trust company should refund the money paid in by stockholders of Bullfrog Rush and prevent Dr. Lyman from selling more stock.

To the brokers, through whom we had sold much of the stock to the public, we telegraphed that we would refund the exact amount paid us by the brokers on delivery back to us of the certificates. We also wired to Governor

Sparks and asked his permission to insert an advertisement in the newspapers over his signature, announcing that the property had proved to be a mining failure and advising the public not to buy any more shares. This pleased the Governor immensely, for he promptly wired back his O. K. wih congratulations over the stand we took.

That night a broadside warning to the public, bearing the signature of Governor John Sparks, and a separate advertisement of the Sullivan Trust Co., offering to refund the money paid for Bullfrog Rush shares, were telegraphed to all the leading newspapers of the East. Next day both of these announcements appeared side by side with the half and full-page advertisements of Dr. Lyman's Union Securities Co. of Goldfield offering Bullfrog Rush for public subscription. The newspapers, peculiarly enough, performed this stunt without a quiver. The public didn't buy any more Bullfrog shares.

The Bullfrog Rush incident cost the Sullivan Trust Co. a little less than $90,000, which was refunded to stockholders, and the additional sum that was expended for advertising our denouncement of the enterprise. Dr. Lyman was stripped of his entire investment in the property. The newspapers lost many thousands of dollars, representing Dr. Lyman's unpaid advertising bills. A number of mining stock brokers also forfeited some money; they were compelled to refund their commissions.

J. C. Weir, the New York mining stock broker, who does business under the firm name of Weir Brothers & Co., had sold in the neighborhood of 100,000 shares of Bullfrog Rush to his clients, and he took violent exception to our decision not to refund an amount in excess of the net price paid to us. He held that his firm ought not to be compelled to disgorge its profits. We stood pat and argued that he ought to be proud to share with us the glory of making good in such an unusual way to stockholders. It was the first time in the history of Western mining promotions that a thing like this had ever been done, and we pointed out to Mr. Weir that it would gain reputation both for himself and the trust company. For a period Mr. Weir carried on an epistolary warfare with the trust company. For nearly two months he refused to yield. Finally, we received a letter from Mr. Weir saying that since we refused to come to his terms he would accept ours, and that he had drawn on us for $4,500, with one lot of 10,000 shares of Bullfrog Rush stock attached. On receipt of the letter I gave instructions to the cashier promptly to honor the draft.

An hour later the cashier reported that the draft had been presented and that an examination of the stock certificates showed that not a single one of them had been sold by the trust company through Mr. Weir's firm, and, in fact, had never been disposed of by the trust company to anybody. A hurried examination of the stock certificate books of the Bullfrog Rush Co., which were in the hands of the company's secretary in Goldfield, a clerk of Dr. Lyman, revealed the fact that a large number of blank certificates had been torn out of the certificate books without any entry appearing on the stubs.

The certificates returned to us by Mr. Weir bore dates of several months prior, and our immediate assumption was that Dr. Lyman, at the very moment when we were marketing the treasury stock under a binding contract which forbade him or any one else to dispose of any Bullfrog Rush stock under any circumstances, was clandestinely getting rid of these shares. Mr. Weir, it appeared, had neglected to segregate Dr. Lyman's certificates from those shipped him by the trust company. Another hypothesis was that those certificates had never been sold at all, but had merely been received from Dr. Lyman to be reforwarded to us in order to claim a refund for what we had never been paid for.

Of course, we returned the draft unpaid. But that didn't end the incident. My partner, Mr. Sullivan, took it upon himself to wire his sentiments to Weir Brothers & Co., as follows: "You are so crooked that if you swallowed a tenpenny nail and vomited, it would come out a corkscrew." That was "Larry's" homely way of expressing his opinion.

Goldfield's year of wind and dust had brightened into the glow of summer. The still breath of August was diffused through the thin, mild air of the high altitude. This thin air, which nearly two years before had prompted a camp wit to comment on the birth of my news bureau to the effect that "the high elevation was ideal for the concoction of the visionary stuff that dreams are made of," appeared unprophetic. There was plenty of concrete evidence of the yellow metal to be seen. Production from the mines was increasing daily and money from speculators was pouring into the camp from every direction.

A mining stock boom of gigantic proportions was brewing. Mohawk of Goldfield, which was incorporated for 1,000,000 shares of the par value of $1 each, and which in the early days went begging at 10 cents a share, was now selling around $2 a share on the San Francisco Stock Exchange, the Goldfield Stock Exchange and the New York Curb. Other Goldfields had advanced in proportion. Combination Fraction was up from 25 cents to $1.15. Silver Pick, which was promoted at 15 cents a share, was selling at 50 cents. Jumbo Extension advanced from 15 to 60. Red Top, which was offered in large blocks at eight cents per share two years before, was selling at $1.

GOLDFIELD GOSSIP

A Monthly Magazine of Nevada Mining

VOL. I   DECEMBER

WARNING!

We stop the press to say that persons who are contemplating coming to Goldfield are earnestly advised to stay away until after the Holidays. By that time we shall have pr[ocured] accommodation for all. This is not the kind of climate to fool with, and people who [are] exposed to its rigors on a winter's night run grave risks. The town is jammed full. There is not a room, a shack, or a cot to [be] had. Certain of our pirate hotels are charging visitors $3.00 for the use of a cot for one night, and packing six of these cots [in] one room. Dozens of people have walked the streets all night, unable to find even this shelter. We are building as fast as we ca[n] but not fast enough for the crowd. Come about the end of January—and come to stay. Be prepared to make Goldfield yo[ur] home. 'Till then content yourselves at home and eat your Christmas dinner in comfort. Enjoy it. It may be your last.—Ed. "Gossip."

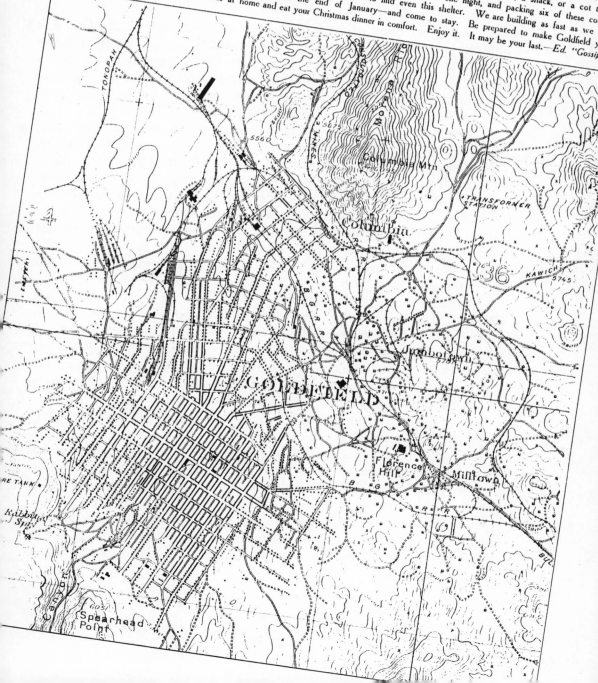

Jumbo advanced from 25 cents to $1.25. Atlanta moved up from 12 to 40. Fifty others, representing prospects, enjoyed proportionate advances.

The Sullivan stocks were right in the swim. Jumping Jack was in hot demand on the San Francisco Stock Exchange and New York Curb at 45 cents, Stray Dog at 70 cents, Indian Camp at 80 cents, and Eagle's Nest at 50 cents. Subscribers to Indian Camp could cash in at a profit of more than 200 per cent.

The country gave indications of going "Goldfield crazy." My Goldfield publicity bureau was working overtime. James Hopper, the noted fiction writer and magazinist, ably assisted by Harry Hedrick and other competent mining reporters, was on the job and doing yeoman service. The news columns of the daily papers of the country teemed with stories of the Goldfield excitement.

People began to flock into the camp in droves. The town was a scene of bustle and life. Motley groups assembled at every corner and discussed the great production being made from the Mohawk and the terrific market advances being chronicled by mining stocks representing all sorts and descriptions of Goldfield properties. Whenever Hayes and Monnette, owners of the Mohawk lease, appeared on the streets, they were followed by a mixed throng of the riffraff of the camp, who hailed them, open-mouthed, as wonders.

The madness of speculation in mining shares in the camp itself was beginning to exceed in its intensity the exciting play at the gaming tables. There was a contagion of excitement even in the open spaces of the street.

At each meeting of the Goldfield Stock Exchange the boardroom was crowded. The sessions were tempestuous. Every step and every hallway leading to the room was jammed with men and women over whose faces all lights and shades of expression flitted. The bidding for mining issues was frantic. Profits mounted high. Everybody seemed to be buying and no one appeared to be willing to sell except at a substantial rise over the last quotations. Castle building and fumes of fancy usurped reason.

Bank deposits were increasing by leaps and bounds. The camp was rapidly becoming drunk with the joy of fortune making.

Manhattan now shone mostly in the reflected glory of Goldfield, but Manhattan stocks were booming. This enabled the Sullivan Trust Co. to dispose of nearly all of its Manhattan securities which had been carried over after the San Francisco catastrophe and to pile up a great reserve of cash.

A big demand was developing for shares in Fairview companies. Nevada Hills of Fairview was selling on the stock exchanges and curbs at $3 per share, or a valuation of $3,000,000 for the mine. Only a few months before

FAIRVIEW
IS THE MECCA OF MINE-MAKERS

STARTING TO SINK—DAY SHIFT ON CLIPPER.

FAIRVIEW NOW THIRD IN PRODUCERS RANK WILL CLIMB IN 1908

Manager of Dromedary Hump Says that District will Soon take Second Place as a Producing Camp

MANY MINES BEING DEVELOPED

Cyclone, Eagle, Eagle's Nest, Dromedary, Reliance, Mizpah and Golden Slipper along with Nevada Hills in Limelight

P. H. McLaughlin, one of the owners of the control of the Dromedary Hump Mines Company and the mine manager of the Company's property in Fairview, was in Reno today. He said: "Fairview's showing, after 15 months of actual development, as good as Goldfield's and Tonopah's was at that time, if not better. Considering that fact, and that Fairview is larger than

it had fallen into Goldfield and Salt Lake hands for $5,000. Fairview Eagle's Nest, for which subscriptions had been accepted at 35 cents per share by the Sullivan Trust Co., was selling at 70 cents on the San Francisco Stock Exchange.

The Sullivan Trust Co. announced the offering of 1,000,000 shares, embracing the entire capitalization of the Fairview Hailstone Mining Co., at 25 cents. The stock was purchased by us at eight cents. We sold out in a week. San Francisco and Salt Lake were the principal buyers, and it was unnecessary even to insert an advertisement offering the stock. The brokers fell over one another to underwrite the offering by telegraph.

PRIZE FIGHTS AND MINING PROMOTION

FOR A FORTNIGHT there was a lull in news of sensatorial gold discoveries, but the approaching Gans-Nelson fight, which was arranged to be held in Goldfield on Labor Day, September 3, furnished sufficient exciting reading matter for the newspapers throughout the land to keep the Goldfield news pot boiling. The Sullivan Trust Co. had guaranteed the promoters of the fight against loss to the extent of $10,000, and other camp interests put up $50,000 more. Gans, the fighter, was without funds to put up his forfeit and make the match, and the Sullivan Trust Co. had also advanced the money for that purpose. Mr. Sullivan became Gans' manager. When Gans arrived in town Mr. Sullivan interviewed him to this effect:

"Gans, if you lose this fight they'll kill you here in Goldfield; they'll think you laid down. I and my friends are going to bet a ton of money on you, and you must win."

Gans promised he would do his best.

"Tex" Rickard and his friends wagered on Nelson. The cashier of the Sullivan Trust Co. was instructed to cover all the money that any one wanted to bet at odds of 10 to 8 and 10 to 7 on Gans, we taking the long end. A sign was hung in the window reading: "A large sum of money has been placed with us to wager on Gans. Nelson money promptly covered inside." Mr. Sullivan was in his glory. Prize fighting suited his tastes better than high finance, and he was as busy as a one armed paper hanger with the itch.

An argument arose about who should referee the fight. "Tex" Rickard nominated George Siler, of Chicago, and Battling Nelson promptly O.K.'d the selection. Mr. Sullivan openly objected. He thought it good strategy. He sent for the newspaper men and gave out an interview in which he declared that Mr. Siler was prejudiced against Gans because he was a Negro, and he did not believe Mr. Siler would give a square deal.

"Rice," whispered Sullivan after the newspaper men left the office, "I am four-flushing about that race prejudice yarn, but it won't do any harm. Siler needs the job. He's broke and I'll make him eat out of my hand before I'll agree to let him referee the fight. They've already invited Siler to come here, and I won't be able to get another referee, but I'll beat them at their own game. When Siler gets here I'll thrash matters out with him and agree to his selection, but first I want him to know who's boss."

Mr. Siler arrived. An hour later he was closeted with Mr. Sullivan in one of the back rooms of the trust company offices. The dialogue which ensued was substantially as follows:

Mr. Siler. You've got me dead wrong, Sullivan. I want to referee this fight, and I want you to withdraw your objections.

Mr. Sullivan. Well, I've heard from sources which I can't tell you anything about that you don't like Gans, and I can't stand for you.

Mr. Siler. I need this fight, and I've come all the way from Chicago in the expectation of refereeing it. I couldn't give Gans the worst of it if I wanted to. He is a clean fighter and I would not have an excuse.

Mr. Sullivan. Gans is a clean fighter, but Nelson isn't; he uses dirty tactics and he is a fouler for fair.

Mr. Siler. If he does any fouling in this fight I'll make him quit or declare him out.

Mr. Sullivan. What guarantee have I got that you won't give Gans the worst of it?

Mr. Siler. Well, I'll tell you, Sullivan, if you withdraw your objections I'll guarantee you that I'll be this fair. If Nelson uses foul tactics, or if he don't, I'll show my fairness to Gans by giving him the benefit of every doubt. Now, will that satisfy you?

Mr. Sullivan. Yes, it'll satisfy me, but, remember, if you don't keep your word you'll have just as much chance of getting out of this town alive as Gans will have if he lays down! You understand?

Mr. Siler. Yes.

On the afternoon of the fight the Sullivan Trust Co. cast accounts and found that it had wagered $45,000 on Gans against a total of $32,500 put up by the followers of Nelson.

Mr. Sullivan, after talking it over with me, had accepted the honorary position of announcer at the ringside. Though not of aristocratic mien, "Larry" was of fine physique, with a bold, bluff countenance, and I felt confident that his cordial manner would appeal to that Far Western assemblage.

Just before the prize fighters entered the ring, "Larry" jumped into the

NELSON'S ARRIVAL IN GOLDFIELD AUG 11th 06
1. BAT. NELSON. 2. NOLAN. 3. RICKARD.
4. GRANT. 5. ELLIOTT. 6. HACKETT. 7. CHRISMAN.

Dubbed the "Fight of the Century," the Joe Gans-"Battling" Nelson Lightweight championship fight which occurred in September 1906 truly put Goldfield in the limelight of publicity.
Its promoter, Tex Rickard is on far left, while Joe Gans, the winner of the 42-round fight, is on far right

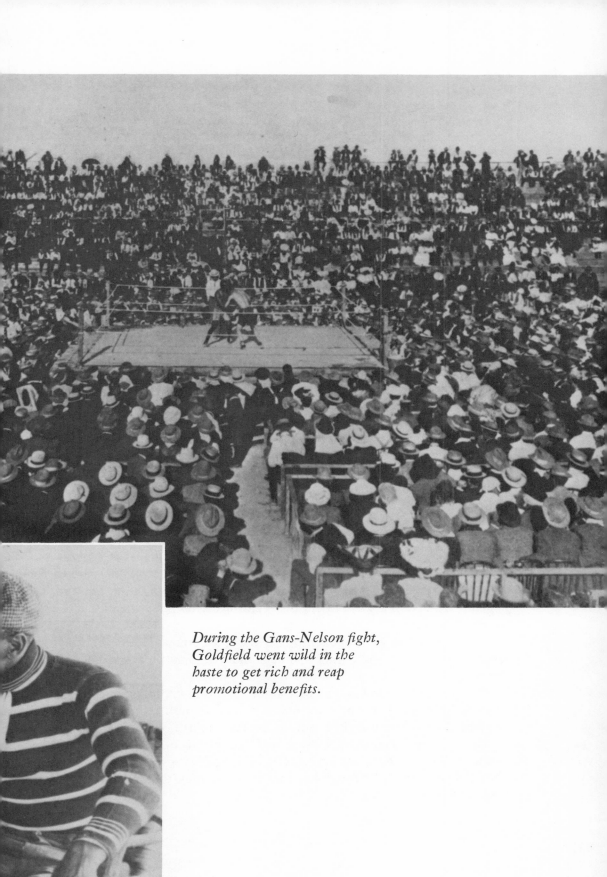

During the Gans-Nelson fight, Goldfield went wild in the haste to get rich and reap promotional benefits.

arena. Standing above the mass of moving heads and holding up both hands, he hailed the great crowd thus:

"Gentlemen, we are assembled in this grand *areno* to witness a square fight. This fight is held under the auspices of 'Tex' Rickard, a man of great *acclumuations—*"

"Larry" did not get much farther. The audience laughed, and then jeered and hooted until it became hoarse. His words were drowned in the tempest of derision. I was informed by friends who were close to the ringside that he went on in the same rambling way for a few minutes more, but I can't testify to that fact from my own knowledge because "acclumuations" and "areno" overcame me and I stopped up my ears.

The fight progressed for twenty rounds or more, when I began to doubt the ability of Gans to win. Mr. Sullivan had a commissioner at the ringside, who, up to this time, had been betting anybody and everybody all the 10 to 6 that was wanted against Nelson. I hailed Mr. Sullivan at the ringside.

"This doesn't look like the cinch for Gans you said it would be," I whispered.

"Wait a minute," Mr. Sullivan replied, "I'll go to Gans' corner as soon as this round is over and find out what's doing with him."

Mr. Sullivan went over to Gans' corner and came back.

"Gans says he can't win this fight, but he won't lose. He's a good ring general and he'll pull us out. Don't bet any more money. I'm going to stay close to the ringside. Watch close."

It was apparent during the next ten rounds that Gans was availing himself of every opportunity to impress upon the audience that Nelson was inclined to use dirty fighting tactics, and soon Nelson was being hooted for foul fighting. Gans, on the other hand, appeared to be fighting fair and like a gentleman. Soon it was evident that Gans had won the sympathy and favor of the audience.

The fight had continued through the 40th round, when Mr. Sullivan again repaired to Gans' corner and held another animated whispered conversation with him.

In the 42nd round Gans all of a sudden went down, rolled over and, holding his hand under his belt, let out a yell of anguish that indicated to the excited multitude that Nelson had fouled him frightfully.

In another instant Mr. Sullivan had clambered into the ring. Confusion reigned. The audience was on its feet. Pushing his fist into the referee's face, Mr. Sullivan cried: "Now, Siler, you saw that foul, didn't you? It's a foul, isn't it? Gans wins, doesn't he?"

All of this happened quick as a flash. Mr. Siler, pale as a ghost, whispered something inaudibly.

Mr. Sullivan, turning to the assemblage and raising both arms to the skies, yelled:

"Gentlemen, the referee declares Gans the winner on a foul!"

The audience acclaimed his decision with salvos of applause. There did not appear to be a man in the crowd who doubted a foul had been committed, although Nelson at once protested his innocence.

Next day Mr. Sullivan told me that in or near the 24th round Gans had broken his wrist and knew he could not win the fight by a knockout. He also said that Gans went down in the 42nd round in order to save the day.

"*I* won that fight," said Sullivan. "I told Gans while he was in his corner after the 40th round that if he lost he would be laying down on his friends, that he had the audience with him, and that it was time to take advantage of Nelson's foul tactics."

This was my first experience in prize fighting, and my last. My sympathies were, however, with the winner. Gans' tactics throughout up to the last round were gentlemanly and those of Nelson unfair. Even the partisans of Nelson who had wagered on him agreed after the fight that the battle put up by the Negro up to the 42nd round was a white man's fight and he was entitled to win.

Nelson had been guilty of foul tactics in almost every round, but the probabilities are that Gans was not disabled by a foul blow in the 42nd round and that he took advantage of the sentiment in his favor, which had been created by his manly battle up to that time, to go down at a psychological moment.

I saw Mr. Siler after the contest, and he appeared pleased that his decision was so well received, but he assured me that if he was invited to referee another bout in any mining camp he would decline the job.

The Sullivan Trust Co., of course, won a big bet on the result, but it lost a bigger one as an outcome of the battle on the very next day. The impression created by announcer Sullivan's attempt to reach lofty flights of eloquence in his speech to the fight audience was bad for the trust company, and it required the use of over $100,000 on the day following to meet the flood of selling orders in Sullivan stock which poured into the San Francisco Stock Exchange.

THE YEAR OF BIG FIGURES

I SOON RECOUPED these stock market losses. At about four o'clock one afternoon, a few days afterwards, a miner who had been at work during the day on the Loftus-Sweeney lease of the Combination Fraction, called at the office of the trust company and asked me to buy 1,000 shares of Combination Fraction stock for him. He divulged to me that just as he was coming off shift he had learned that a prodigious strike of high-grade ore had been made at depth. Combination Fraction had closed that afternoon on the San Francisco Stock Exchange with sales at $1.15. I went out on the street and proceeded to buy all the Combination Fraction in sight. In half an hour I had corralled about 60,000 shares at an average of $1.30. An hour later the owners of the lease obtained the information on which I was working, and by eight o'clock that night, when the Goldfield Stock Exchange began its evening session, the price had jumped to $1.85. Within a week thereafter the price sky-rocketed to $3.75, and at this figure I took profits of nearly $150,000. Had I held on a little longer I could have doubled that profit, for Combination Fraction a few weeks later sold at higher than $6.

The Combination Fraction strike was followed by a number of others, and the boom gathered force. By October, Goldfield Silver Pick had advanced to $1 per share, up 600 per cent. Goldfield Red Top was selling at $2, Jumbo at $2, and Mohawk at $5, showing profits of from 2,000 to 5,000 per cent. Others had gained proportionately. In fact, there were over twenty Goldfield securities listed on the exchange that showed the public a stock market profit of anywhere from 100 per cent to 5,000 per cent.

Mining machinery of every description was being shipped into camp, and for half a mile around the Combination mine the landscape of assembled gallows frames resembled a great producing oil field. There were signs of mining activity everywhere. For four miles east of the Combination mine and six miles south every inch of ground had been located. Claims situated miles away from the productive area were changing hands hourly at high figures.

The Sullivan stocks kept pace in the markets with the other booming securities, and it was plain that the trust company was riding on a tidal wave of success. Our profits exceeded $1,500,000 at this period, and we were just eight months old.

In a single fortnight the Sullivan Trust Co. promoted the Lou Dillon Goldfield Mining Co. at 25 cents per share, a valuation of $250,000 for the property, which cost $50,000; and the Silver Pick Extension, which cost $25,000, at the same figure, netting several hundred thousand dollars' profit on these two transactions. Options to purchase Lou Dillion and Silver Pick Extension, which were situated within 500 feet of the Combination mine, had been in possession of the Sullivan Trust Co. for months, and had increased in value to such an extent that on the day the subscriptions were opened in Goldfield for Lou Dillon at 25 cents per share, a prospector named Phoenix, who had received $50,000 from the Sullivan Trust Co. for the entire property, subscribed for 100,000 shares, or a tenth interest in the enterprise, paying $25,000 therefor.

It was the rule of the Sullivan Trust Co. to open subscriptions in Goldfield on the day its advertising copy left the camp by mail for the East. Newspaper publishers were always instructed to publish the advertisements, which were generally of the full page variety, on the day following receipt. In the case of Lou Dillon it became necessary to telegraph all newspapers east of Chicago not to publish the advertisement because of oversubscription before the copy reached them, and in the case of Silver Pick Extension the orders to publish the advertisements were canceled by telegraph before the mail carrying the copy reached Kansas City. San Francisco, Los Angeles and Salt Lake subscribed for 50 per cent of the entire offering of Lou Dillon and Silver Pick Extension, and Goldfield for 25 per cent. As a matter of fact, had we desired, we could have sold the entire offerings in Goldfield, Tonopah and Reno without inserting any advertisements, so great was the excitement in the state itself.

At this period the combined monthly payrolls of the mining companies promoted by the Sullivan Trust Co. totaled in excess of $50,000, and excellent progress was being made in opening up the properties.

It was early autumn in Goldfield, warm, dry and dusty, and never a cloud in the sky. I was at my desk eighteen hours a day, and liked my job. Things were coming our way.

The Sullivan Trust Co. was in politics. Mr. Sullivan was popular with the miners, and Governor Sparks was a large asset of the trust company because he had been allowing the use of his name as president of all the mining companies promoted by it. Nevertheless, when the state election approached, the Governor had no money for campaign expenses. He telegraphed the trust company from Carson:

"I will not stand for renomination."

We replied, "You are certain to be elected, and you will be renominated by acclamation if you accept."

"I won't run unless you guarantee my election," he telegraphed.

We answered, "We guarantee."

The Governor was renominated by the Democrats. The Republicans placed in nomination J. F. Mitchell, a mining engineer and mine owner, who was very popular among mine operators.

There were thousands of miners domiciled in Goldfield. The Western Federation of Miners dominated.

"Sullivan," I said, "isn't it a certainty that the miners will vote the Democratic ticket because Mitchell has been put forward by the mine owners? Is it necessary to spend any money with the Western Federation?"

"Not a dollar!" replied Mr. Sullivan. "There's a meeting of the executive committee tomorrow. I'm going to be around when they meet. Without spending a cent I'll bring home the bacon. Watch me!"

Sullivan reported to me the next day that he had succeeded in his mission.

"I didn't attend the meeting," he said, "but I did see the main 'squeeze.' He told me that a contribution to the Miner's Hospital would be gratefully accepted, but that even that was not necessary, and that Sparks would win in a walk."

The only campaign money advanced by the Sullivan Trust Co. was given to Mr. Sullivan to go to Reno. He asked for $1,000, and he used it in conducting open house on the first floor of the Golden Hotel, meeting people and greeting them. Reno appeared to be a Republican stronghold, and Mr. Sullivan, by baiting the Catholics against the Protestants, succeeded in holding down the Republican majority to an extent that was wofully insufficient to overcome the Democratic majority rolled up in Goldfield with the aid of the miners. Governor Sparks was reëlected by a handsome majority. Had the occasion demanded it, we would have "tapped a barrel." But it was not necessary.

THE STORY OF GOLDFIELD CONSOLIDATED

RUMORS WERE RIFE in Goldfield of a merger of mammoth proportions which was said to be on the tapis. Great as were the gold discoveries in camp, they did not justify the terrific advances being chronicled in the stock market, and it was apparent that something extraordinary must be hatching to justify the market's action.

George Wingfield, who had enjoyed a meteoric career, rising within five years from a faro dealer in Tonopah to the ownership of control in the Mohawk and many other mining companies and to part ownership of the leading Goldfield bank, John S. Cook & Co., which was then credited with having $7,000,000 on deposit, was said to be engineering the deal. The names of the properties were not given, nor the figures. It occurred to me that in any merger that was made the Jumbo and Red Top, because of their central location, must be included. I sought out Charles D. Taylor, who with his brother, H. L. Taylor, and Capt. J. B. Menardi, owned the control of these properties. He asked $2.50 per share for his stock and that of his partners—all or none. Mr. Taylor had walked into the camp as a prospector. Most of his nights were spent at the gambling tables, and he was reported to be an easy mark for the professionals. His losses were constant and heavy. I put Mr. Sullivan on his trail. Mr. Sullivan reported to me that Mr. Wingfield was hobnobbing with Mr. Taylor.

"Get an option on these properties from Taylor and be quick," I told Mr. Sullivan.

Next morning I met Mr. Sullivan. He held in his hands 20,000 shares of Jumbo, selling at $1.75 per share on the Goldfield Stock Exchange.

"I won it in a poker game last night with Taylor and Wingfield," he said. "I have an oral option on the property good for three days at $2.50, but if you leave it to me, I'll win these properties from him playing cards."

I did not see Mr. Sullivan again for a week. Next I heard of him he had "fallen off the water wagon" and was reported to be celebrating the event in Tonopah. While Mr. Sullivan was kidding himself about his poker playing ability, Mr. Wingfield had come to terms with Mr. Taylor and had bought the control of Jumbo and Red Top at an average of $2.10 per share. That explained Mr. Sullivan's lapse. However, I blamed myself. Mr. Sullivan was no match for Mr. Wingfield. In any game from stud poker to marketing mining stock Mr. Wingfield can outwit, outmaneuver and outgeneral a hundred like "Larry."

Both companies had been capitalized for 1,000,000 shares. The sale required that a fortune be paid over. Mr. Wingfield paid a small sum down, and Mr. Taylor placed the stock of both of these companies in escrow in the John S. Cook & Co. bank, the balance to be paid a month later.

The purchase of control of the Jumbo and Red Top by the firm of Wingfield and Nixon signaled the beginning of a stock market campaign for higher prices that stands unprecedented for audacity and intensity in the history of mining stock speculation in this country since the great boom of the Comstock lode in 1871-1872.

The market for all listed Goldfield stocks was made to boil and sizzle day in and day out until Jumbo and Red Top had been ballooned from $2 to $5 per share, Laguna from 40 cents to $2, Goldfield Mining from 50 cents to $2, and Mohawk from $5 to $20. Within three weeks the advance in market price of the issued capitalization of this quintet alone represented the difference between $8,000,000 and $26,500,000.

A few days before top prices were reached, it was officially announced that the merger of Mohawk, Red Top, Jumbo, Goldfield Mining and Laguna

Plenty of fight money was at stake, and the L. M. Sullivan Company, the front for George Graham Rice, assisted winner Joe Gans in every way possible.

into the Goldfield Consolidated Mines Co. had been made on the basis of $20 for each outstanding share of Mohawk, $5 for Red Top, $5 for Jumbo, $2 for Goldfield Mining, and $2 for Laguna. It was also given out that the promoters, Wingfield and Nixon, had allotted themselves $2,500,000 in stock of the merged companies as a promoters' fee. Right on top of this came an announcement that the Combination mine had been turned into the merger for $4,000,000 in cash and stock, and it was learned that go-betweens had made a profit of $1,000,000 on the deal by securing an option on the property for $3,000,000.

In short, a merger was put through of properties and stocks, the issued capitalization of which was selling in already inflated markets on the day the merger was conceived for $11,000,000, at a valuation of $33,000,000, and in additon the promoters received a $2,500,000 bonus. Had the properties been merged on the basis of their selling prices three weeks prior, the equivalent value of the 3,500,000 shares of merger stock would have been a fraction above $3. As it stood, under the ballooning process, the market value was $10, which was the par.

At the time of the merger these were the conditions that ruled at the mines: The Mohawk, appraised at $20,000,000, had produced under lease in the neighborhood of $8,000,000, of which less than $2,000,000 had found its way into the treasury of the Mohawk Mining Co., the balance going to the leasers. The leasers had high-graded the property to a fare-you-well, and less than $1,000,000 worth of high-grade remained in sight, although it was conceded on every side that the leasers had not attempted, nor were they able during the period of their leasehold, to block out systematically and put into sight all of the ore in the mine. Large, but indefinite, prospective value therefore attached to Mohawk in addition to the tonnage in sight.

The Laguna, for which $2,000,000 had been paid in stock, did not have a pound of ore in sight, and had cost Wingfield and Nixon less than $100,000. Goldfield Mining, scene of a sensational production during the early days of the camp, appraised at $2,000,000 more, had fizzled out as a producer.

Jumbo, taken in for $5,000,000, for a year previous had produced little or no ore, most of the time being exhausted by the management in sinking a deep shaft, and it had less than $500,000 in sight. Red Top, valued at another $5,000,000, had in excess of $2,000,000 worth of medium grade ore blocked out.

Wingfield and Nixon were also heavily interested in Columbia Mountain, Sandstorm, Blue Bull, Crackerjack, Red Hills, Oro, Booth, Milltown, Kendall, May Queen, and other Goldfield stocks. No sooner did the five stocks forming the merger begin to show such startling market advances than the ballooning tendency manifested itself in Wingfield and Nixon's miscellaneous list, and all of them showed phenomenal gains. Soon the entire list of Goldfield, Tonopah, Manhattan, Bullfrog, and other Nevada mining securities listed on the San Francisco Stock Exchange and traded in on the exchanges and curbs of the country, felt the force of the terrific rises, and sympathetically they skyrocketed to unheard-of levels.

To convey an idea as to how far the prices of these stocks were moved up beyond their intrinsic worth, as a result of the ballooning process of the merger, I give some comparisons.

Columbia Mountain sold during the boom at above $1.50; it is now selling at five cents. Blue Bull, Crackerjack, Oro, Booth, Red Hills, Milltown, Kendall, Conqueror, Hibernia, Ethel, Kewanas, Sandstorm and May Queen sold at an average of 75 cents during the boom; they are now selling at an average of less than five cents. A hundred other Goldfield securities, which were in eager demand at the zenith of the spectacular movement at prices ranging from 50 cents to $2.50 can now be purchased at from one to five cents per share, while many others that were hopefully bought by an over wrought public at all sorts of figures are now not quoted at all.

AT THE HEIGHT OF THE FRENZY

THE DIFFERENCE BETWEEN the market price of listed Nevada stocks on November 15, 1906, and that of today is in excess of $200,000,000. A fair estimate of the public's real money loss in the listed division is $150,000,000.

Nor was this all of the damage that was done. When excitement in Goldfield's listed stocks reached a frenzy, wildcatters operating from the cities got into harness, and within three months in the neighborhood of 2,000 companies, owning in most instances properties situated miles from the proved zone in Goldfield, or in unproved camps near Goldfield, were foisted on the public for $150,000,000 more.

The fact that Mohawk, which in the early days of Goldfield could have been purchased at 10 cents, had advanced to $20 and had shown purchasers a profit of 20,000 per cent; that Laguna had advanced in less than two years from 15 cents to $2; that Jumbo and Red Top, selling at $5, could have been purchased a year or two before at around 10 cents; that Goldfield Mining, which had in the early days been peddled around the camp at 15 cents, had moved up to $2, etc., gave wildcatters an argument that was convincing to gulls in every town and hamlet in the Union. And the harvest was immense. Not one of the 2,000 wildcats has made good, and every invested dollar has been lost.

It will be noted from the reckoning as given that about as much money was lost in the listed stocks of the camps as in the unlisted cats and dogs.

As a matter of fact, veteran mining stock buyers, in camp and out of the camp, lost as much hard cash as did the unsophisticated. San Francisco, which owes its opulence of years gone by to successful mining endeavor, was probably hit as hard as any other city in the Union. San Francisco thought it knew the game, and it confined its operations to the stocks listed on the exchange where the Comstocks are traded in. But San Francisco did not know the inside of the merger deal as it is now known to every schoolboy in Nevada.

The operation on the inside was this: Wingfield and Nixon owned the John S. Cook & Co. bank in Goldfield, and they owned the control of nearly a score of mining companies which were of little account as well as having acquired the control of the biggest mine in camp. During the height of the boom, which they engineered to swing the merger, they disposed of millions of shares of an indiscriminate lot of companies, and used the many millions of proceeds to take over Jumbo, Red Top and their outstanding contracts in Mohawk and other integrals of the merger. They likewise were able during the ballooning process to dispose of much Mohawk at from $15 to $20, much Jumbo at from $4 to $5, much Red Top at from $4 to $5, that cost them very considerably less than this, and in this way were enabled to finance their deal to a finish.

I have just pointed out that in order to accomplish the merger it was necessary that the market in all Goldfield securities, in which the promoters were interested, be stimulated in order to enable unloading by the insiders before some of the very large payments became due. This being accomplished, and the payments having been made, the promoters sought to establish a market for merger shares at or around par. In order to accomplish this the Goldfield bank, in which the promoters were heavily interested, stimulated speculation and managed to spread a feeling of security by announcing its willingness to loan from 60 to 80 per cent par on merger shares. All Goldfield fell for this, and the camp went broke as a result.

Within eighteen months thereafter Goldfield Consolidated sold down to $3.50 in the markets, and margin traders and borrowers who had put up the stock as collateral to purchase more were butchered. Loans were foreclosed by the bank as rapidly as margins were exhausted. The carnage was awful.

It must be evident that Wingfield and Nixon, both of whom became multi-millionaires as the result of their mining stock operations in Goldfield, were directly and indirectly important factors in the loss by the public of $300,000,000, as set forth above. It is admitted that less than $7,000,000

THIS IS FOR YOU
MR. MINING INVESTOR

THE NEVADA MINING NEWS is the great unshackled, absolutely independent and fearless mining and financial newspaper published in the State; the one which tells the truth regardless of whom it hurts or helps.

¶ The NEVADA MINING NEWS stands neither in **the glare of self-made halos,** nor under the shadow of market machinations.

¶ It has only one axe to grind—it will, come what may or who will, protect the interests of buyers of Nevada mining shares. **Within the range of its vision the big culprit is no better than the little one.** Against both it aims to protect the investor.

¶ Can you, a buyer of Nevada mining stocks, afford to do without the NEVADA MINING NEWS?

Do you want the truth?
Do you want facts about men?

? about mines
about prospects
about districts
about stocks

The Nevada Mining News
Gives Facts Only

¶ For subscriptions forwarded now, with $5, the regular annual price, enclosed, we will credit subscribers up to March 15, 1909. That means **FIFTY-EIGHT ISSUES of the Livest, Most Fearless, Most Independent, Most Truthful mining newspaper ever printed at the price of 52 issues.**

SUBSCRIBE NOW
GET RIGHT ON NEVADA

ADDRESS
NEVADA MINING NEWS
140 VIRGINIA ST. :: RENO, NEVADA

George Graham Rice was the Editor

worth of ore had been developed as a reserve at the time $35,000,000 worth of stock in the merger was issued and a market manufactured to dispose of the stock at this fictitious price level. It is not of particular interest that Goldfield Consolidated, by reason of sensationally rich mine developments at depth, has since given promise of returning to stockholders an amount almost equal to par for their shares, and that it now appears that those who were able to weather the intervening declines may in the end be out only the interest on their money.

This fact stands out: although Goldfield Consolidated owned at the outset a bonanza gold mine, stockholders had just two chances. They could break even or lose—break even on their investment if the mine made good in a sensational way, which was a big gamble at the time, or lose if the mine didn't. They could not win.

Mr. Nixon was a United States Senator from Nevada. He was also president of the Nixon National Bank of Reno, Nevada. He held both of these positions at the time the merger was made, and it was largely because of Mr. Nixon's political and financial position that the daring ballooning market operations, which were staged as a curtain raiser for the merger, proved so successful.

In the *Nevada Mining News* of May 25, 1907, circulation 28,000, an interview appeared with United States Senator Nixon of Nevada, vouched for as follows:

> The manuscript of the interview was submitted to, and approved by, the Senator. Unchanged by one jot or tittle, it is printed just as it came from his hands. Even now the Senator holds a carbon of the original manuscript and may brand us with it if we have broken the faith we pledged.

I quote from the Senator's interview, as it appeared in that issue of the *Nevada Mining News*:

> "What do you estimate the ultimate earnings of Goldfield Consolidated will be?" was asked.
>
> "Consolidated will be a bigger producer, I should say, three or four years from now than it will be one year from now," Senator Nixon replied, "and I believe I am conservative when I say that the property will be eventually earning $1,000,000 net monthly."
>
> "Then, as an investment, the stock is easily a $20 stock?"
>
> "That is a minimum estimate of its future value, I should say," was the response.

As to that interview, Mr. Nixon said that within three or four years (the time limit is up), $20 would be a minimum price for the shares. They touched $10 only once since then, or one-half of his estimate. Shortly after the interview was given they sold down as low at $3.50. Recently the market quotation was $4.

He said, further, that the mines would ultimately earn at the rate of $1,000,000 a month. This statement also has fallen far short of fulfillment.

Soon after George S. Nixon, as president of the Goldfield Consolidated Co., gave out this interview for public consumption he, according to his own later admissions, disposed of all of his holdings, and at an average price, it is believed, of less than $8 a share.

This is only a superficial rendering of the big event in Goldfield's history, but it is sufficient to furnish an example of the effect of get-rich-quick influences that radiate from high places and separate the public from millions upon millions, without being called to account.

The dear American public has been falling for this kind of insidious brand of get-rich-quick dope for years. It is being gulled into losing millions through its fetish worship of promoters with millions, who are really the get-rich-quicks of the day that are very dangerous.

Greenwater, a rich man's camp, in which the public sank $30,000,000 during three months that marked the zenith of the Goldfield boom, is another case in point where a confiding investing public followed a deceiving light and was led to ruthless slaughter.

OFFERING HIGH BIDS ON HIGH GRADE MINING STOCK ON DOOR MINING STOCK EXCHANGE IN GOLDFIELD, NEVADA.

4. The Greenwater Fiasco

WHEN THE EXCITEMENT was at fever heat in Goldfield over the stupendous rises in market value of Goldfield securities which were being chronicled hourly, news came to town of the successful flotation in New York of the Greenwater & Death Valley Mining Co. The capitalization was 3,000,000 shares of the par value of $1 each. The stock had been underwritten at $1 a share by New York and Pittsburgh Stock Exchange houses, had been listed on the New York Curb, and had climbed to around $5.50, or a valuation for the property of $16,500,000. Among the officers of this company were M. R. Ward, brother-in-law of Charles M. Schwab; T. L. Oddie, now Governor of Nevada, and Malcomb Macdonald, later president of the Nevada First National Bank of Tonopah.

Greenwater is situated about 150 miles south of Goldfield, across the state line in California. No one ever went to or fro without passing through Goldfield. If there was a Greenwater boom, how was it that we in Goldfield, who were in touch with all Nevada mining affairs, did not know about it? Goldfield promoters soon began to give attention. Shortly they caught the infection. A stampede from Goldfield into Greenwater ensued. In fact, people flocked to Greenwater from every direction. A bunch of Tonopah money getters, headed by the indomitable Malcolm Macdonald, were grabbing the money on Greenwaters in New York, and Goldfield was not in the play.

The reports that came from Greenwater as a result of the first stampede from Goldfield were of doubtful variety. Greenwater & Death Valley was described as a raw prospect not worth over 10 cents per share. Goldfield people shook their heads. There was no gainsaying the fact, however, that Greenwater & Death Valley appeared to be a giant success in the Eastern stock markets. Charles M. Schwab was reported to be behind the flotation of Greenwater & Death Valley. Montgomery-Shoshone and Tonopah Exten-

Greenwater Broker
(INCORPORATED)

ARTHUR KUNZE, President
E. T. GRADY, Vice President

CAPITAL,

Mines === Stocks === Investm

COPPER

Gre

This is the initial issue of COPPER TALK. It is a review and bonanza mining camps of Southern Nevada and Calif and the wonder copper camp---Greater Greenwater---e freely, frankly and and as concisely as it can. Its basis operation and exploitation. Its motive is the extension the cause of mining investment education. It will be want it and is submitted for what it is worth.

THE GENERAL SITUATION.

THE general situation is fairly expressed by the two words—Incomparable Prosperity. Incomparable because we have no standard to measure it by. The volume of business is unparalleled. It has broken all records in every branch and channel of business enterprises—industrial, commercial and financial. The production total of the farm, the plantation and the mine is past the highest high water mark, mills, factories and railroads are compelled to Yea

The Greenwater 1907 boom

sion, two other Schwab enterprises, were selling at hundreds of per cent profit in the stock markets. The fact that Mr. Schwab was interested in the camp was an argument that appealed with great force to Nevada promoters, for the fraternity had learned to attach just as much significance to having a market as to having a mine before commencing promotion operations.

The Sullivan Trust Co. not having had a failure of any kind on the market, I hesitated to commit the trust company to any issue in the new camp. Not to be entirely out of it, however, I sent our engineer, "Jack" Campbell, into the district to report on all the properties.

News came thick from the New York market as to the success of the Greenwaters in the East. Furnace Creek Copper Company, originally promoted by "Patsy" Clark of Spokane at 25 cents per share, with a million share capitalization, was reported to be getting the benefit of Mr. Clark's personal market handling on the New York Curb, and the shares soon reached a high quotation of $5.50. John W. Gates had been let in by "Patsy" at around 50 cents and was reported to have unloaded 400,000 shares at all sorts of prices from $1 up to $5.50, and down again.

On the heels of this advance came word of the successful promotion of the United Greenwater Co., with C. S. Minzesheimer & Co., members of the New York Stock Exchange, acting as fiscal agents for the company. The promoters were named as Malcolm Macdonald, Donald B. Gillies and Charles M. Schwab. J. C. Weir, the New York mining stock broker, who was conducting through the mails a nation-wide market letter campaign in favor of Greenwater, was reported to have sold 150,000 or 200,000 shares at the subscription price of $1. The offering was said to have been oversubscribed twice. The price then shot up to $2.50 on the New York Curb. The market boiled.

Philadelphia was reported to be Greenwater-mad. When United Greenwater had reached $1.50 on its way up and Greenwater & Death Valley had passed the $4 point, the Schwab crowd announced the formation of the Greenwater Copper Mines & Smelters Co. to consolidate the Greenwater & Death Valley and United Greenwater companies. This new parent company was capitalized for $25,000,000, with 5,000,000 shares of the par value of $5 each, and the East was reported to be eating up the new stock "blood raw." The president of this company was Charles R. Miller, who was president of the Tonopah & Goldfield Railroad Co., and the vice-president was M. R. Ward, the redoubtable brother-in-law of Charles M. Schwab. The directorate included Mr. Schwab; John W. Brock, who represented Philadelphia interests on the directorate of the very successful Tonopah Mining Co.; Malcolm

Macdonald, the champion lemon peddler of Nevada; Frank Keith, general manager of the Tonopah Mining Co., and others. It was a "swell" directorate.

It was learned that the stock of the new company had been underwritten by New York Stock Exchange houses, principally those with Philadelphia and Pittsburgh branches where the Schwab crowd was influential, at $1.80 per share, and that large blocks were being sold to the public at up to $3.25 on the New York Curb, a valuation for the properties of more than $16,000,000.

GETTING INTO THE GAME

THE BIRTH OF THE $25,000,000 merger, to take in two properties that had not yet matriculated even in the baby mine class and were actually suspected at the outset by the mining men in Goldfield to be wildcats, was the signal for an outpouring in quick succession of Greenwater promotions from all centers, of which the annals of the industry in this country chronicle no counterpart.

At the height of the boom there was promoted out of Los Angeles and New York the Furnace Creek Consolidated Copper Co., with a capitalization of $5,000,000.

From Butte, home of the copper mining industry, the Furnace Creek Extension Copper Mining Co. was promoted, with a capitalization of $5,000,000, and also the Butte & Greenwater, capitalized for $1,500,000. Malcolm Macdonald, the hero of Montgomery-Shoshone at Bullfrog, hailed from Butte. He it was who interested the Schwab crowd in Greenwater, as he did in Tonopah and Bullfrog.

"Patsy" Clark, the noted mine operator of Spokane, having prospered marketwise with his Furnace Creek Copper Co., promptly headed a new one, the Furnace Valley Copper Co., with a capitalization of $6,250,000. These shares were listed on the Spokane, Butte and Los Angeles Stock Exchanges, but did not appear on the New York Curb.

A San Francisco crowd of brokers and stock market operators organized the Greenwater Bimetallic Copper Co. "They let her go Gallagher" with a capitalization of $1,000,000.

The C. M. Sumner Investment Securities Co. of Denver opened subscriptions for the Greenwater-Death Valley Copper Co. (The title of this company was a play on the name of the Greenwater & Death Valley Copper Co.)

Tonopah citizens, not to be outdone, sallied forth with the Greenwater Calumet incorporated for $1,500,000. Hon. T. L. Oddie, later Governor of Nevada, then of Tonopah, and his brother, C. M. Oddie, followed the lead and headed the Greenwater Arcturus Copper Mining Co., with a capitalization of $3,000,000.

The Consolidated Greenwater Copper Co. was fed to the hungry public out of a Pittsburgh trough, with general offices in the Keystone Bank Building, and with a high class Tonopah crowd on the directorate. Eugene Howell, cashier of the Tonopah Banking Corporation, of which United States Senator Nixon was president, was treasurer. John A. Kirby, of Salt Lake City, until recently associated with George Wingfield in the ownership of Nevada Hills, was president.

Arthur Kunze, who had sold the control of the Greenwater & Death Valley Copper Co. to Malcolm Macdonald, who in turn had interested the Schwab coterie in the organization, put out a new one called the Greenwater Copper Mining Co., with a capitalization of $5,000,000.

H. T. Bragdon, formerly president of the Goldfield Mining Co., which is one of the integrals of the Goldfield Consolidated, headed the Greenwater Black Jack Copper Mining Co., with a capitalization of $1,000,000.

ALL THE COPPER IN THE WORLD

UNITED STATES SENATOR George S. Nixon of Nevada lent his name, along with H. H. Clark, William Bayley and H. J. Woollacott, as a director of the Greenwater Furnace Creek Copper Co., with a capitalization of $1,500,000. The prospectus of this company announced that the ores were "melaconite, azurite, chalcocite, and occasionally chrysocolla, averaging 18 to 36 per cent (copper) tenor."

"Taking the lowest percentage of ore reported by the company," says Horace Stevens in the *Copper Handbook* of 1908, "and the company's own figures as to the size of its ore bodies, the first 100 feet in depth on this wonderful property would carry upward of 20,000,000 tons of refined copper, worth, at 13 cents per pound, the comparatively trifling sum of five billion, two hundred million dollars."

Mr. Stevens goes on: "The fact that a Major is manager of this company, and a United States Senator is vice-president, will prove a great consolation to the shareholders. It is indeed lamentable to note that this magnificent mine, which carries, according to the company's own statements, more copper than all the developed copper mines of the world, is idle, and present office address a mystery."

Donald Mackenzie, of Goldfield, promoter of the successful Frances-Mohawk Mining & Leasing Co. at Goldfield, which netted over $1,500,000 from Mohawk ores, and distributed all of 20 per cent of this amount to stockholders in the shape of dividends, pushed out the Greenwater Red Boy Copper Co. and the Greenwater Saratoga Copper Co., with a capitalization of

$1,000,000 each. Thomas B. Rickey, president of the State Bank & Trust Co. of Goldfield, Tonopah and Carson City, was president of both of these companies, and J. L. "God-Bless-You" Lindsey, cashier of the State Bank & Trust Co., was treasurer.

Greenwater Consolidated, Greenwater Copper, Furnace Creek Oxide Copper, Greenwater Black Oxide Copper, Greenwater California Copper, Greenwater Polaris Copper, Greenwater Pay Copper, Pittsburgh and Greenwater Copper, Greenwater Copper Range, Greenwater Ely Consolidated, Greenwater Sunset, New York & Greenwater, Greenwater Etna, Greenwater Superior, Greenwater Victor, Greenwater Ibex, Greenwater Vindicator, Greenwater Prospectors', Greenwater El Captain, Greenwater & Death Valley Extension, Greenwater Copper Queen, Greenwater Helmet, Tonopah Greenwater, Furnace Creek Gold & Copper, and Greenwater Willow Creek were the names of a score of others with capitalizations ranging all the way from $1,000,000 to $5,000,000 each.

Among these the Greenwater Willow Creek Copper Co. boasted of the fanciest directorate. George A. Bartlett, Nevada's lone Congressman, was president, and Richard Sutro, then head of the world known New York banking house of Sutro Bros. & Co., was advertising as first vice-president. Henry E. Epstine, the popular Tonopah broker, was second vice-president, and Alonzo Tripp, general manager of the Tonopah & Goldfield Railroad, was a director.

Did I fall for Greenwater? Yes, and at the eleventh hour. On the half-hearted recommendation of the trust company's engineer, "Jack" Campbell, the L. M. Sullivan Trust Co. paid $125,000 for a property in Greenwater that boasted of two ten-foot holes. On two sides it adjoined the property of the Furnace Creek Copper Co., the original location in the camp. Our engineer reported that if "Patsy" Clark's Furnace Creek Copper Co. shares of which were selling in the market at a valuation of $5,500,000 for the property, had any ore, we certainly could not miss it. No matter which way the veins trended, our ground must be as good as "Patsy's," because the identical vein formation passed through both properties.

The Sullivan Trust Co. thereupon incorporated the Furnace Creek South Extension Copper Co. to operate the property. The capitalization was 1,250,000 shares of the par value of $1, of which 500,000 shares were placed in the treasury of the company to be sold for purposes of mine development.

New York Stock Exchange houses having the call as purveyors of this particular line of goods, the Sullivan Trust Co. tendered the selling agency of Furnace Creek South Extension treasury stock to E. A. Manice & Co.,

Though short on ore, Greenwater had plenty of publicity, and the Chuck-Walla's mailing list extended to the 45 states. Rice himself estimated that the American public sank $30 million in Greenwater in less than four months in 1907. "Yet the suckers . . . were crying for more."

The Death Valley Chuck=Walla

A MAGAZINE FOR MEN

Volume 1, Number 6.
Greenwater, California
April the 1st, 1907.

¶ Published on the desert at the brink of Death Valley. Mixing the dope, cool from the mountains and hot from the desert, and withal putting out a concoction with which you can do as you damn please as soon as you have paid for it. ❧ ❧ ❧ ❧   PRICE, TEN CENTS members of the New York Stock Exchange, whose officers are located in the same building in New York as J. P. Morgan & Co. We offered for public subscription 100,000 shares of treasury stock at par, $1, through E. A. Manice & Co., and this firm advertised the offering in New York newspapers over their own signature. The Sullivan Trust Co. paid the bills.

THE COLLAPSE OF GREENWATER

THE OFFERING TURNED OUT to be a "bloomer," the first the Sullivan Trust Co. had met with. E. A. Manice & Co. did not dispose of as many as 30,000 shares. Neither did the stock offered later by the Sullivan Trust Co. through brokers in other cities sell freely. Just at the moment when we announced our offering of Furnace Creek South Extension the Greenwater boom began to crack.

Oscar Adams Turner, who promoted the Tonopah Mining Co. of Nevada, which has paid $8,000,000 in dividends on a capitalization of $1,000,000, is responsible for the early bursting of the bubble. Mr. Turner had invested in the Greenwater camp on the reports of an engineer. He organized the Greenwater Central Copper Co. Then he decided that it was advisable for him to take a look at the property for himself. He visited Greenwater. Two hours after arriving in camp he sent a telegram to Philadelphia reading substantially as follows:

> Stop offering Greenwater Central. Make no more payments on the property. Do not use my name any further. There is nothing here.

The tenor of the message leaked out. Indiscriminate selling ensued by a noted bank crowd in Philadelphia who were loaded up with Greenwaters. Others followed suit. The market became sick.

At the first sign of a market setback inquiries began to pour into Nevada from all over the East, and noted copper experts from Montana, Arizona, California and other points came piling into the Greenwater camp to examine the properties. Soon a chorus of adverse opinion found its way into every financial center. Market values crumbled as rapidly as they had risen. Paper fortunes evaporated in thin air.

I make a conservative statement when I say that the American public sank fully $30,000,000 in Greenwater in less than four months.

Not all of the Greenwater promotions were over-subscribed—not half, not a quarter—and the American public may well congratulate itself that the boom busted when only approximately $30,000,000 had passed into the pockets of the promoters.

What of the camp? It exists no more. All mine development work ceased long ago. There are green-stained carbonates on the surface, but there are no copper ore bodies. The "mines" have been dismantled of their machinery and other equipment, and not even a lone watchman remains to point out to the desert wayfarer the spot on which was reared the monumental mining stock swindle of the century. Every dollar invested by the public is lost. The dry, hot winds of the sand-swept desert now chant the requiem.

Fix the responsibility here if you can. The job is not easy. Let me attempt it. The buccaneers who took Greenwater & Death Valley down to New York and allowed the public to subscribe for it with the name of Charles M. Schwab as a lure, at a valuation for the property of more than $3,000,000, and then ballooned the price on the curb until the shares sold at a valuation of $16,500,000 for the property, without an assured mining success in sight in the entire camp—these men, in my opinion, were criminally responsible. They have never been called to account.

Members of the New York Stock Exchange who aided and abetted them by lending their names to the transaction, and Charles M. Schwab, who permitted the use of his name and that of his brother-in-law, are morally responsible. Not for an instant do I entertain the thought that the Stock Exchange crowd and Mr. Schwab realized that the mines of the company were absolutely valueless, but I do maintain that men of their standing and prestige have opportunities which men of small caliber do not enjoy and that their conduct for this reason was reprehensible to an extreme.

THE SHAME AND THE BLAME

I CITE THE INSTANCE of the Sullivan Trust Co. falling for Greenwater, after hesitating about embarking on the enterprise for weeks, and I am convinced that others feel the same way. The Sullivan Trust Co. did not touch a Greenwater property until its clients and its clientele among the brokers throughout the Union had burned up the wires with requests for a Greenwater promotion, and when it did finally fall it lost its own money, the only other sufferers being a handful of investors who at the tail end of the boom subscribed for a comparatively small block of treasury stock.

Not all of the promoters fell innocently, however. There were half-baked promoters and mining stock brokers in almost every city in the Union who had witnessed the enhancement in values during the Goldfield boom, and whose palms had itched for the "long green" that for so long came the way of men on the ground. These, at the first signal that the Greenwater boom was on, with Charles M. Schwab in the saddle, lost no time in annexing ground in the district with the single view of incorporating companies and retailing the stock to the public at thousands of per cent profit.

The Greenwater mining boom fiasco stands in a class by itself as an example of mining stock pitfalls. The only Greenwater stock which at this time has a market quotation is Greenwater Mines & Smelters, which reflects the true state of the public mind regarding all Greenwaters by actually selling at a valuation of less than the amount of money in the company's treasury—six cents per share on an outstanding issue of 3,000,000 shares—there being $189,000 in the treasury along with an I.O.U. of C. S. Minzesheimer & Co., the busted New York Stock Exchange house, for $71,000, of which the company will realize 27 cents on the dollar through the receiver.

Greenwater in 1968

5. The Great Goldfield Smash

IT WAS EARLY IN November, 1906. Indian summer held Goldfield in its soft embrace. Nature wore that golden livery which one always associates with the idea of abundance. The mines of the district were being gutted of their treasures at the rate of $1,000,000 a month. Under the high pressure of the short-term leasing system new high records of production were being made. The population was 15,000. Bank deposits totaled $15,000,000. Real estate on Main Street commanded $1,000 a front foot. The streets were full of people. Every one had money.

In years gone by men died of thirst on that very spot. Three years before there were no mines and the population numbered only a corporal's guard. The transformation was complete. Within three years the dreams of the lusty trail blazers, who had braved the perils of the desert to locate the district, had become a towering reality. The camp, which two years before was dubbed by financial writers of the press as a "raw prospect" and a "haven for wildcatters and gamblers," had developed bonanza proportions. The early boast of Goldfield's press bureau, that Goldfield would prove to be the greatest gold camp in the United States, was an accomplished fact.

Listed Goldfield mining issues showed an enhancement in the markets of nearly $150,000,000. Stocks of neighboring camps had increased in market value $50,000,000 more. The camp rode complacently on the crest of the big boom, than which history chronicles no greater since the famous old days of Mackay, Fair, Flood and O'Brien on the Comstock. There was no premonition that a climax must be reached in climbing values at some period, and that a collapse might be near.

Goldfield Consolidated shares were selling on the exchanges at above par, $10, or at a market valuation of more than $36,000,000 for the issued capitalization of the company. You could have bought all of the properties of

BY-LAWS

ARTICLE I.
Title.

The title of this Association shall be "Rhyolite Mining Stock Exchange."

ARTICLE II.
Charter and Regular.

Section 1. There shall be two classes of Membership members, Charter and Regular. Both members shall consist of those who possess full membership, and are entitled to all privileges of the Exchange.

ARTICLE III.
Governing Committee.

Section 1. The whole government of the Exchange shall be vested in a mittee composed of eleven bers, elected in the manner vided.

Sec. 2. The members of Committee shall be elected Exchange members, and the ing the largest number of v fice shall be declared elected

Sec. 3. The members o Committee shall be divided i as nearly equal as possible, first class expiring in one ye class in two years, and of three years.

Sec. 4. At each annual e change there shall be chose Governing Committee to fil casioned by the outgoing office for the three years members to fill any vacan the other classes, for the Their term of office shall b

Sec. 5. The Governing have the right to remove for any good cause, of wh

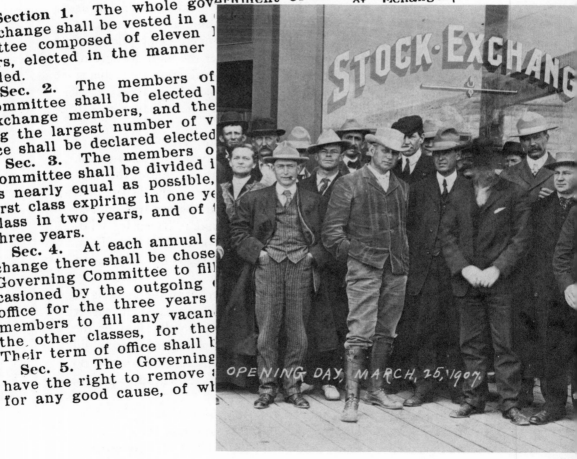

During 1906-1907 Rhyolite in the Bullfrog District became a center of stock promotion, according to Rice

A. H. TEN BROECK
President

A. B. ACORN
General Manager

J. HOWARD BARNARD
Sec'y and Treas.

Original Bullfrog Extension Mining Company

Capital Stock, $1,000,000
Fully Paid and Non-Assessable

1,000,000 Shares
Par Value, $1.00 Each

THE FIRST LOCATION made in the Bullfrog Mining District was that of the Original Bullfrog; the second that of the Hillside and the Hillside No. 1, the property of this corporation. Our mining expert, Mr. A. B. Acorn, after sampling all the principal claims that have given the district its wonderful reputation, selected these as the most desirable, and says today that had no more development work been done on the other locations, he would still consider, from all the surrounding and surface indications, the Hillsides the most promising claims in the district.

¶ Our Superintendent reports that the rich vein of the Original Bullfrog crosses our ground. We are sinking a shaft on the contact of this vein with our independent lead, an 8-foot ledge of very promising ore, which has been traced the entire length of both claims, a distance of 3000 feet. Assays from the surface of this ledge ran from $1.60 to $34 per ton, by far the best showing made in the vicinity previous to development.

¶ From the experience of other mines, we have every assurance of rich ore deposits extending over an unusual proportion of these claims. We have remarkable facilities for working them on account of the formation of the ground, being able to run tunnels on the ledge in both directions, giving 400 and 600 feet of backs, respectively.

¶ Some of our principal stockholders have incorporated a syndicate for the purchase and development of good prospects, and the placing of the same on the market as a whole, or, if deemed more favorable, incorporating and handling them for the benefit of investors. We shall be pleased to correspond with parties who have claims to sell, with prospectors in the field, or with all who propose to work on these lines. Our connections with capitalists and Eastern investors enable us to reach mining investors, and should aid us materially in placing properties in the right hands. Orders for stock and all correspondence relating to mining matters may be addressed to

J. HOWARD BARNARD, Secretary, 175-178 Crocker Building, San Francisco, Cal.

this company for less than $150,000 when the camp was first located. A score of leases were operating the Consolidated's properties. The leases were soon to expire. Much market capital was made of the fact that the company would presently come into its own.

More than 175 stocks of Goldfield and nearby camps were listed on the exchanges and curbs. All of these were selling at sensational prices and enjoyed a swimming market. The successful merging by Wingfield and Nixon of the principal producing properties of Goldfield at a $36,000,000 valuation, more than four times the value of the known ore reserves, stimulated the whole list.

Columbia Mountain, promoted by the mergerers of Goldfield Consolidated, but excluded from the merger because not contiguous to the other integrals and because it had no ore, had been ballooned to $1.35 per share on a million share capitalization, and stood firm in the market regardless of the fact that it was still only an unpromising prospect. The issued stock of a dozen other companies in control of the promoters of the merger was selling at an aggregate value of many millions more. The most despised "pup" in this particular group was Milltown, of not even prospective value; yet it easily commanded a per-share price that gave the property a market valuation of $400,000.

Silver Pick, capitalized for 1,000,000 shares of the par value of $1 each, had scored an uninterrupted advance from 15 cents to $2.65 a share without a pound of ore being found on the property. The market price did not waver.

Kewanas, another million share company, was in big demand at $2.25 per share, a valuation of $2,250,000 for the property and an advance of 2,250 per cent over the promotion price. Kewanas's gain was also made despite the fact that mine developments had failed to open up pay ore in commercial quantities. Eight months earlier the entire acreage had been offered to me for $35,000 and I had refused to buy.

Goldfield Daisy, promoted by Frank Horton, a faro dealer in George Wingfield's Tonopah gambling joint, had been ballooned from 15 cents to $6 a share on a capitalization of 1,500,000 shares. It had never earned a dollar for stockholders, but was actually selling in the open market at a valuation of $9,000,000. The price showed no sign of weakening.

Combination Fraction, owning a few acres of ground, which was promoted at 20 cents a share on a capitalization of 1,000,000 shares, had risen rapidly, because of ore discoveries and contiguity to the Mohawk, to $8.50 a share. Stockholders gave no sign of a tendency to unload.

Great Bend, situated in the Diamondfield section of the Goldfield district, four miles from the productive zone, had been carried up from 10 cents a share to $2.50 without a mine being opened up, establishing a market valuation for the property of $2,500,000.

These are but a few of the more striking instances of price appreciations. All of these stocks, excepting Goldfield Consolidated, are now selling for a few pennies per share each, the average not being so much as ten cents. There were over a hundred other Goldfield stocks that also enjoyed spectacular market careers, on which it is now impossible to get any quotation at all.

THE RISE OF WINGFIELD AND NIXON

ANY ONE IN GOLDFIELD who was willing to admit that stocks were selling too high at the time was decried as a knocker. You could borrow freely on all listed Goldfield stocks at John S. Cook & Co.'s bank, owned by the promoters of the Goldfield Consolidated, and the men of the camp for that reason felt that there must be concrete value behind nearly all of them. Brokers in Eastern cities reported that few of their customers were willing to take profits even at the prices to which stocks had been skyrocketed. Most mining stock brokers of the cities had knocked the stocks of the camp in the early days before the advance. At this stage, when prices had reached undreamed-of levels, the brokers did not advise their customers that values had been worked up far beyond intrinsic worth. Indeed, they actually waxed enthusiastic in their recommendations to buy. Every one was a bull.

Sessions of the Goldfield Stock Exchange reflected the extent of the craze. Outside of the exchange the stridulous, whooping, screeching, detonating voices of the brokers that kept carrying the market up at each session could be heard half a block away. Later, did you find your way into the crowded board room, the half-crazed manner in which the notebooks, arms, fists, index fingers, hats and heads tossed and swayed approached in frenzy a scene of violence to which madness might at once be the consummation and the curse.

George Wingfield and his partner, George S. Nixon, were the heroes of the hour. Less than five years before, Mr. Wingfield had come into Tonopah with a stake of $150, supplied by Mr. Nixon, whose home was in Winnemucca, Nevada. Mr. Wingfield had formerly been an impecunious cowboy gambler. Born in the backwoods of Arkansas, and later of Oregon, he hailed from Golconda, Nevada. Mr. Nixon, at the time he staked Mr. Wingfield and until his election as a United States Senator in 1904, was known as the "state agent" of the Southern Pacific Company for Nevada, having succeeded on the job the notorious "Black" Wallace, who for many years handled

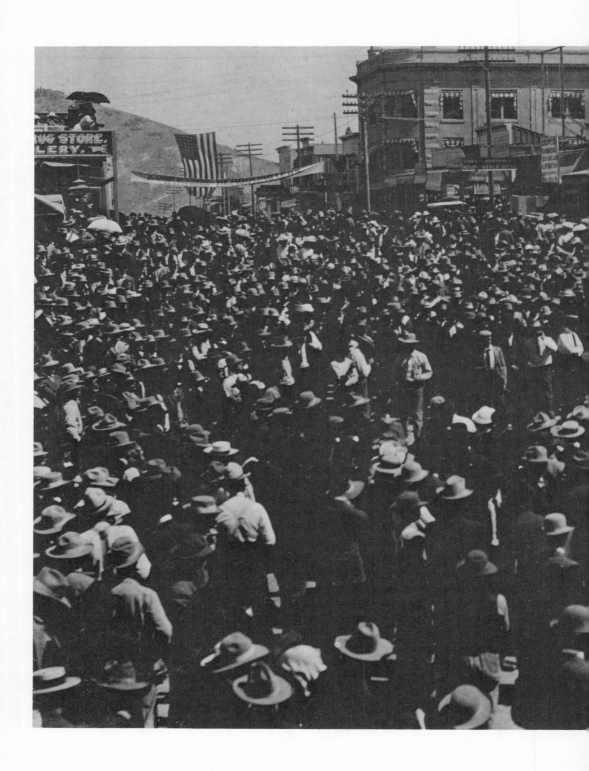

Rock drilling contest in downtown Goldfield, 1907

the "yellow dog" fund for the Huntington régime when franchises were hard to get and legislatures had to be bought. Mr. Nixon was also president of a bank in Winnemucca, which was a way station on the Southern Pacific Railroad.

Mr. Wingfield had signalized his money-getting prowess by running Mr. Nixon's $150 into $1,000,000 as principal owner of the Tonopah Club, the biggest gambling house in Tonopah, and later parleying the money for himself and partner into ownership of control of the merged $36,000,000 Goldfield Consolidated, which was their corporate creation.

Mr. Wingfield was said to be behind the market. He was looked upon as boss of the mining partnership, and Mr. Nixon as a circumstance. Mr. Wingfield was a conspicuous figure at nearly all the sessions of the Goldfield Stock Exchange, of which he was a member. In the early evenings, when informal sessions were held on the curb, he could also be seen in the thick of the tumult. He was on the job at all hours.

At that time Mr. Wingfield was about thirty years old. Of stinted, meager frame, his was the extreme pallor that denoted ill health, years of hardship, or vicious habits. His eyes were watery, his look vacillating. Uncouth, cold of manner, and taciturn of disposition, he was the last man whom an observer would readily imagine to be the possessor of abilities of a superior order. In and around the camp he was noted for secretiveness. He was rated a cool, calculating, selfish, sure-thing gambler man of affairs—the kind who uses the backstairs, never trusts anybody, is willing to wait a long time to accomplish a set purpose, keeps his mouth closed, and does not allow trifling scruples to stand in the way of final encompassment. Among stud poker players who patronized gaming tables in Tonopah, Goldfield and Bullfrog, he was famed for a half-cunning expression of countenance which deceived his opponents into believing he was bluffing when he wasn't. In card games he was usually a consistent winner.

His partner, George S. Nixon, looked the part of the dapper little Winnemucca bank manager and confidential state agent of the Southern Pacific that he was before becoming Senator. He was considerably below middle weight, and above middle girth at that part of his anatomy which a political enemy once described as seat of his thoughts and the tabernacle of his aspirations. His steel-gray eyes were absolutely without expression. Newly rich, his money and his Southern Pacific connections had gained him a toga, but he did not carry himself like a man upon whom the honors had been thrust. Around Goldfield he strutted with the pride and gravity of a Spanish grandee.

The pair were in control of the mine, bank and market situation. Brokers, bank men and officers of mining companies waited upon them and did their bidding. At night, in the Montezuma Club, where leading citizens were wont to congregate, Mr. Wingfield would on occasion ostentatiously offer to wager that Goldfield Consolidated "would sell at $15 before $9," etc. Men with money who had flocked to the camp from every direction listened in rapt attention. At a later hour they secretly wired the news to their friends in the East. Next morning the market would reflect more public buying and still higher prices. Goldfield itself was blindly following the lead of the twain. It was indeed easier for these men to mark prices up than to put them down.

THE WINNINGS OF A TENDERFOOT

WHAT ABOUT ME? Where did I stand and what was my position at this conjucture? Did I have foresight? Did I realize that stocks were selling at much higher prices than were warranted by intrinsic worth and speculative value? Was not the fact that the mergerers and waterers of Goldfield Consolidated were in command of the mine, market and bank situation sufficient to make me suspect that possibly the cards might be stacked and that maybe cards were being dealt from the bottom of the deck? Was I, in fact, wise to the exact situation and did I realize a smash was bound to ensue? 'Tis a pity hindsight were not foresight, for only in that event could I laurel wreath myself.

I had been on the ground for more than two years. In reality I was still a tenderfoot. My experiences had been unique—all on the constructive side. I had mastered the first rudiments of the game, but only the first. Intrinsic value didn't figure as the only item in my conception of the worth of a Goldfield mining issue. The millionaires of the camp were not miners by profession and their judgment of the value of any mining property would not have influenced a Guggenheim, a Ryan or a Rothschild to extend so much as $4 on the development of any piece of likely mineral ground. Goldfield was a poor man's camp. And it was making good despite the croakings of school trained engineers who had turned the district down in the early days, as they did Tonopah.

At this period I was living frugally. I never touched a card. I worked at my desk on an average of sixteen hours a day, including Sunday, and I never relaxed. Although I had arrived in the camp broke, had I been offered $2,000,000 for my half interest in the L. M. Sullivan Trust Co. I think I should have refused it.

I liked my job. The leaven of my environment appealed directly to my perceptions. I was saturated with the traditions of Western mining luck and also with the optimism of my sturdy neighbors. These men had stood their ground in the early period of the camp's days of trial and tribulation. They had triumphed like their forebears on the Comstock, just as did the hardy pioneers of Leadville and Cripple Creek and as their brethren of Tonopah did. Their influence over me was unbounded. I relished the work, anyhow. As a matter of fact, I had little use for money except for the purpose of business. And never a suggestion came to me that it was time for a "clean up."

The L. M. Sullivan Trust Co., of which I was vice-president and general manager, was doing remarkably well. The stocks of the mining company were listed on the San Francisco Stock Exchange and New York Curb and showed a market appreciation of $3,000,000 above the promotion prices. Indian Camp, promoted at 25 cents, was selling freely at $1.30. Jumping Jack, for which subscriptions were originally accepted at 25 cents, was in hot demand at 62 cents. Stray Dog, sold to the public originally at 45 cents, was active around 85 cents. Lou Dillon, put out less than a month before at 25 cents, had worked its way up to 64 cents. Silver Pick Extension, which was oversubscribed at 25 cents and commanded 35 cents two hours after we announced that subscriptions were closed, was selling on the exchanges and curbs of the country at 49 cents. Eagle's Nest Fairview, which original subscribers got into at 35 cents, was very much wanted at 65 cents. Fairview Hailstone, floated at 25 cents, was in constant demand at 40 cents. Governor John Sparks was now president of all these companies.

You could have sold blocks of the Sullivan stocks at these profit-making prices on any of the mining exchanges and curb markets of the country without reducing the price a cent, so constant was the public demand and so broad was the market. With the exception of Bullfrog Rush, for which the Sullivan Trust Co. had refunded the money to subscribers when the mine under development proved to be a lemon, every promotion of the trust company showed investors a handsome stock market profit. In the aggregate the promotion price of the seven Sullivan mining companies figured $2,000,000 for the entire capitalization. The market price of these was now $5,000,000, or an average of 150 per cent.

It was a record to be proud of, and I was proud of it, not alone because I was vice-president and general manager of the trust company, but also because a firm of expert accountants, recommended by the American National Bank of San Francisco to examine the books of the trust company, had reported that our assets were $3,000,000 in excess of liabilities, all of which had been gathered in about ten months' time. About $1,000,000 of this represented promotion profits. The remainder was earned by the appreciation in price of mining securities carried or accumulated through the boom.

It was the common boast of the camp that George Wingfield had parleyed or pyramided $1,000,000 which represented the profits of his gambling place in Tonopah, into ownership of control along with his partner Nixon, of the $36,000,000 Goldfield Consolidated. As heretofore related, I had experienced a lot of hard luck in missing by a hair's breadth, ownership of the Hayes-Monnette lease on the Mohawk and the Nevada Hills mine, which would have increased our profits $8,000,000 more, but I felicitated myself that I had done very well by pyramiding $2,500 into a half interest in a flourishing $3,000,000 trust company. I was vain enough to believe that my achievement was as unique as that of Mr. Wingfield, because he had had the influence of a United States Senator and the money deposited in a chain of newly established banks in Goldfield, Tonopah and other points to aid him in his operations. Against this I had not only been compelled to rely on my own resources, but was actually required to combat the work of blackmailers who from time to time attempted to levy tribute. On my failure to "come through" (I never did) they rarely hesitated to take a malevolent smash in print at the Sullivan Trust Co., because in years gone by its active head happened to have had a very youthful past, even though they knew that past was no longer his and he had passed it like milestones on the way.

I AM LANDED HIGH AND DRY

THE NEVADA STATE ELECTION took place in November. The Democratic ticket, headed by "Honest" John Sparks for Governor and Denver S. Dickerson for Lieutenant-Governor, was victorious. The Republican ticket, headed by J. F. Mitchell, a mining promoter and engineer, backed by United States Senator Nixon, the Republican political boss, suffered humiliating defeat.

Denver S. Dickerson was the candidate of the labor unions. During a former labor war in Cripple Creek Mr. Dickerson had been confined in the bull-pen when the governor intervened to quell the labor riots there. Goldfield miners to a man very naturally voted for him. Governor Sparks had accepted the renomination at the urgent request of the L. M. Sullivan Trust Co., and his victory, as well as the complexion of the ticket, was credited largely to the activities in politics of the trust company.

The trust company, while not a banking institution in the sense that it accepted deposits of cash from citizens of the town, having confined its operations to the financing of mining enterprises, loomed large on the political and business horizon because of its increasing financial and political power.

George Nixon

George Wingfield

The trust company carried all of its money in banks that were not affiliated with the Wingfield-Nixon confederacy and worked at cross purposes with it in this particular, too.

The Wingfield-Nixon crowd had pyramided a gambling house in Tonopah and a little one-horse bank in Winnemucca into ownership of control of the $36,000,000 Goldfield Consolidated; into ownership of John S. Cook & Co.'s bank in Goldfield, which was credited with deposits aggregating $8,000,000; into a new bank in Tonopah, known as the Tonopah Banking Corporation, and into a newly formed bank in Reno, called the Nixon National. In politics it had succeeded in seating Mr. Nixon in the United States Senate, placing at his command the Federal patronage which goes with that exalted office.

The confederacy was reaching out. In Goldfield it had overcome such strong banking opposition as the Nye & Ormsby County Bank and the State Bank & Trust Co., both of which were in business before John S. Cook & Co. were dreamed of. It had accomplished this by loaning large sums of money to Goldfield brokers and other citizens on mining stocks of the camp at a time when this class of securities was not so readily accepted by the other banks as good collateral. In Tonopah the newly established Nixon bank, known as the Tonopah Banking Corporation, was making gradual headway against both the Nye & Ormsby and the State Bank & Trust Co., which still carried about 75 per cent of the business of that camp. In Reno the Nixon National found it hard to compete with such old institutions as the Bank of Nevada, the Washoe County Bank and the Farmers & Merchants National, but rumors were already in the air that the Nixon bank was soon to buy out and consolidate with the powerful Bank of Nevada.

In Goldfield the power of the confederacy was strongest in all lines except politics. There it already had its grasp on the throat of the mining and financial business of the camp, and through the out of town draft collection department of its bank held its finger on the pulse of the mining share markets. Its sore spot was politics.

Wingfield and Nixon's market operations were clouded in mystery. No one knew exactly where they stood. Brokers in Goldfield and San Francisco, who had compared notes, were convinced that the two had unloaded many millions of shares of the small companies not included in the merger, and had raked in not less than $10,000,000 during the boom as the result of this selling. The disposal of huge blocks of stock by Wingfield and Nixon, however, was not interpreted as meaning that stocks were selling too high. The general idea prevailed that the proceeds were used to enable the confederacy to finance its stock purchases in the integral companies that were turned over in the making of the merger and to finance its new chain of banks.

About the middle of November the market for Goldfield securities took a turn for the bad. Prices gave indication of having reached a stopping place. Goldfield promoters began to complain that they were compelled to lend strong support to the market because of selling from many quarters that could not be explained. There was much market pressure. In a few days the market became unsteady, then soft, then wobbly again. In camp Wingfield and Nixon were reported still bullish.

The securities of the Sullivan Trust Co. were under attack in all markets. Salt Lake and San Francisco were reported to be spilling stock. Great blocks were being thrown over. I gave support in a jiffy.

There was no surcease. Within ten days I was forced to throw all of a million dollars behind the market to hold it.

This didn't faze me. I was getting stock certificates for the money, and I believed they were worth the price. But I was puzzled to determine what it was all about.

THE BEGINNING OF THE RAID

SOON IT WAS REPORTED to me that Senator Nixon was advising people at all points who held Sullivan stocks, or knew of anyone who held them, to unload. From San Francisco came word that a clique of brokers was operating for the decline.

On the following Monday the market on the San Francisco Stock Exchange opened strong and buoyant, and it looked for a moment as if the selling movement had collapsed. I felt relieved.

My 'phone bell rang. A stock broker of Tonopah called me on the long distance.

"Offer you 10,000 Lou Dillon at 48," he said. "Do you want them?" Lou Dillion was a Sullivan stock that had been promoted at 25; 48 was now a point under the market, however.

"We'll take 'em," I said. "What's the matter?"

"Rumored up here that your books are under inspection by the post office department. You have had five new men on your books for the past few weeks, and some one has spread a story here that Nixon has sicked the Government on to you."

I denied it, of course. The five men in question were the experts who had been sent up from San Francisco by the firm of accountants recommended to us by the American National Bank, and they were at our own behest. The story was a raw canard.

Throughout the day the Sullivan Trust Co. was called upon to stand behind the San Francisco market and take in nearly all of the big blocks of Sullivan stocks owned in the campus of Tonopah and Manhattan. Before our denials could reach the sellers the damage had been done. And it took $250,000 a day for four days to hold the market against this fresh onslaught.

Color had been lent to the wild rumors about a postal investigation by the fact that an attack had been made on me in the columns of the *Denver Mining Record* a year before. Rumor said the dose was going to be repeated. In the early days of the camp, when I was at the head of the Goldfield-Tonopah Advertising Agency, I had represented the *Record* in Goldfield. As its agent I had secured advertising contracts for it which netted my agency in the neighborhood of $10,000 a year in commissions. The owners of the newspaper conceived the idea that I was making too much money on a commission basis and sent Wing B. Allen, formerly of Salt Lake, to the scene to take my place. Mr. Allen worked for smaller pay. He wanted me to divide my commission on standing business, and I refused. The publishers took Mr. Allen's part. As a result I withdrew all the advertising from the columns of the *Denver Mining Record* for which my agency had been responsible, and the *Record* was never able to regain the lost ground.

A short time before the raid on our stocks began Mr. Allen had been arrested in Goldfield on a warrant sworn out by L. M. Sullivan, tried before Judge Bell on the charge of extortion and bound over to the Grand Jury. At the hearing before Judge Bell the Sullivan Trust Co. submitted evidence that Mr. Allen had threatened if we did not give his paper a slice of the promotion advertising of the Sullivan Trust Co., that the *Denver Mining Record* would commence to attack me personally in its columns, and, because of my early past, would do the trust company serious damage.

At the hearing despatches were submitted which were filed at the Goldfield office of the Western Union Telegram Co. by Mr. Allen, in which he had informed his paper that it had better proceed with the attack, because neither Mr. Sullivan nor myself gave indication of yielding. At the hearing, under oath and in a crowded courtroom, I openly denounced Mr. Allen and his newspaper as blackmailers of the very vilest type, and so did Mr. Sullivan. Judge Bell, on the submission by the Western Union of Mr. Allen's despatches to his paper, promptly held him for the Grand Jury.

On the advice of former Governor Thomas of Colorado, to whom the Sullivan Trust Co. paid a retainer as counsel, and who later became chief

Weepah boomed in the late 1920's

counsel for the Goldfield Consolidated, I employed Christopher C. Clay of Denver to commence suit against the owners of the *Denver Mining Record*. As a result I secured from them a settlement by which they agreed not to mention my name again in their paper. I was harassed at the time, or I would not have compromised. The stuff printed by the *Record*, which has been rehashed by every blackmailer who ever attempted to levy on me, was about two-tenths true and eight-tenths false. It was a literal copy of an anonymous publication put out by a set of blackmailers who had tried to circulate it years before in New York when I was head of the Maxim & Gay Co. I had spent thousands of dollars to run down the authorship then, but without avail. The lawyers had succeeded in seizing thousands of copies of the publication, and had made an arrest, but they failed to prove authorship of the screed and ownership of the paper, and the culprits therefore were not punished. In Denver when Mr. Clay applied for criminal warrants, he was asked first to furnish proof of authorship, which was impossible for us, the articles having been unsigned.

SOME PERTINENT PERSONALITIES

THE SAME STUFF has recently appeared without signature in a Goldfield paper which originally came into possession of George Wingfield through foreclosure proceedings, and in a Reno paper controlled by Senator Nixon, who owns a large slice of the paper's mortgage. It has also appeared in other papers friendly to Wingfield and Nixon. Tens of thousands of copies of the Goldfield publication containing the anonymous libel have been circulated.

Other newspapers have reproduced the libelous stuff, some innocently and some for sordid reasons, but of this more later. My career is fraught with instances of recourse by enemies to blackmail and attempted blackmail. If I should undertake to tabulate the cases where men and interests, ranging from impecunious newspaper reporters to financial newspaper publishers and mining stock brokers and market operators who, from the background, publish market letters or furnish the capital for mining publications, have attempted to levy tribute or to club me into submission by the use of so vile a weapon, I should be compelled to write a big book on the subject.

And right here I should like to place myself on record to the effect that seemingly the principal shortcoming that has marked my mining-financial career has been that I had a youthful past—a past which during the last decade has never been taken into serious consideration by men who have held close business relations with me, but which, of course, is a thorn in the sides of men and interests whose bidding I have failed to obey.

I defy any man to cite a single instance where I was guilty of crookedness in a mining transaction or a business transaction of any kind in my entire career as a promoter. I have been fearless—too much so. I have been a rabid enthusiast. I have tried to build. I have given quarter, but have never taken any. I have been honest. Were I really dishonest, I could have prevented every publication of an attack of consequence on me by lending myself in advance to the base purposes of my traducers, and I would have millions now for having compromised with them. It is heaven's own truth that in nine cases out of ten, when I have been attacked in print, the motive of the attacking party has been base and the facts have been so distorted or misrepresented that the fabric was a lie. Nor has the cruelty of the operation stayed any one's hand.

At the very moment in Goldfield when I knew that the *Denver Mining Record* would not assault the Sullivan Trust Co. again because of the settlement of the libel suit by my lawyers out of court, fresh rumors were spread that the *Record* was getting ready for another attack and that tens of thousands of copies of that newspaper were to be circulated. But you can't stop a rumor by the declaration of the truth, and the Sullivan Trust Co. decided that it would be unwise to make a denial in print, for by so doing it would communicate to all stockholders the news that the Sullivan stocks were actually under attack and thus cause more frightened selling.

Sight drafts from brokers in New York, Chicago, Salt Lake and San Francisco, drawn on the Sullivan Trust Co., with large bundles of Sullivan stocks attached, were pouring into our office through the local banks for presentation. John S. Cook & Co. made a specialty of this department of banking, and most of the drafts on us were cleared through the Wingfield-Nixon bank. It was reported to me that Senator Nixon was openly discussing the enormous volume of stocks coming in on us and was questioning our ability to stem the tide. As a strategic measure, the Sullivan Trust Co. decided to cross sales on the San Francisco Stock Exchange so that it might ship out of the camp, through the banks, large blocks of stock with draft attached against San Francisco brokers and thus convey to the minds of local bankers that we were selling large blocks of stock as well as buying them. The volume of the cross trades caused some talk in San Francisco, and was magnified by brokers operating for the decline.

THE TIME WHEN MONEY TALKS

SOME OF OUR BROKERS in San Francisco now demanded an independent bank guaranty that the drafts on us would be honored. We asked for a line of credit at the State Bank & Trust Co. It was promptly given. As fast as the brokers asked for a guaranty, the State Bank & Trust Co. telegraphed them formally that it would honor our paper to the extent of $20,000 or $30,000 in every case. To protect the bank and in order to be able to borrow a large sum of money, should we need it in the event of another selling movement starting in, we deposited stocks of a market value of $1,500,000 with the State Bank & Trust Co., which signed a paper that this collateral was to stand against loans for any amount which the State Bank & Trust Co. might make to us on open account.

A few days later we borrowed from the bank $300,000 in cash, and it was agreed that should we need $300,000 more on the same collateral, it would be promptly placed at our disposal. We did not yet need the money, but I realized the desirability of assembling cash in an exigency such as that. Nor was this an unusual proceeding. There was a time during the Manhattan boom when the overdraft of the Sullivan Trust Co. in the Nye & Ormsby County Bank was $695,000. The bank held against this overdraft Sullivan stocks at the promotion price. Nearly all of these stocks at that early period were as yet unlisted.

The idea of withdrawing support and letting the market go to smash did not occur to me at all. As already stated, I believed the stocks were worth the money. But that was not the chief reason for my stubborn market position. I took great pride in the fact that every listed stock of the Sullivan Trust Co. showed a big profit to stockholders. I considered the greatest asset of the trust company to be, not its money, but its prestige, and I entertained big ideas as to a future I had mapped out for the corporation. I did not suspect that an organized campaign was on to destroy us and that the dominant interests of the camp were reaching out for everything in sight. Nor did I have any use for money for hoarding purposes. The only thing that seriously nettled me was the fact that the Sullivan Trust Co. had been compelled to turn borrower.

Before the first selling movement started in, our assets were $3,000,000 more than our liabilities. But this $3,000,000 was not all cash. In fact, it was represented in part by stocks which we had purchased in the market with the idea that they were good stocks to own and would show the trust company a big profit, as they had. We could have cleaned up $3,000,000 in cash, but we had not done so. Now, within a month, all of our available cash had been put into fresh lines of our own securities, we had been compelled to sell other lines out, and the corporation was a borrower. I was stubborn—too stubborn for a man who boasted of so little experience in such a big game. It was a pet belief of mine that obstacles create character. I was in the heat of a battle and fighting my way against tremendous odds. I rather liked the sensation.

Another dominant trait which, deep down, has in recent years been the keynote of my actions is the fact that my philosophy teaches me that you can't down the truth, that a lie can't live, and that justice will be finally done. Had I always put the accent on the finally and mixed with my philosophy a little dope to the effect that while justice is always *finally* triumphant, injustice is often victorious *for a while*, I might have fared better.

Previously I stated that "Wall Street deals for suckers" and that "thinkers who think they know, but don't" are the suckers for which Wall Street casts its net. I also stated that Wall Street promoters realized that "a little knowledge is a dangerous thing" and that this "little knowledge" leads astray this particular kind of sucker. In falling in Goldfield for the philosophy that "justice is always triumphant in the end," by swallowing it whole, and in making no allowance for the fact that justice is sometimes tardy, even though it does prevail in the end, I here decorate myself with a medal as a top-notcher in the *sucker* class—in the academic sense—which I have described, and which is the usual sense in which I use the term "sucker."

Again the selling ceased, and it looked as if the Sullivan Trust Co. would be compelled to wait only for a general turn in the market to relieve itself of money pressure by disposing of some of the large blocks of stock it had accumulated during the periods of heavy liquidation.

CLOUDS IN THE WESTERN SKY

A NEW BLACK CLOUD showed itself on the horizon. A labor war was threatening in Goldfield. It was very apparent, from the conduct of George Wingfield, that he was baiting the miners, and it appeared to be the general opinion of the people of Goldfield that he was trying to precipitate trouble. The miners had asked for higher wages. The Sullivan Trust Co., which was operating seven properties with a monthly payroll of $50,000, was the first to express a willingness to grant the terms. Wingfield and Nixon refused. The miners asked for arbitration. It was refused. The mines were then shut down for a few days and the terms of the leases were extended.

Heavy selling in all Goldfield stocks took place during the shutdown. Rumors could now be heard on every side that Wingfield and Nixon were dumping overboard big blocks of stock. Could it be possible that they themselves were scuttling the ship that had given them such glorious passage?

Again the Sullivan Trust Co. was called upon to stand behind the market.

The Great Goldfield Smash

Soon a cry of distress was heard in the camp from investors and stock brokers who had overloaded themselves with securities and who were in debt to the banks to the extent of millions, with stock of the camp put up as collateral. Inquiry revealed the fact that all Goldfield and Tonopah banks were overloaded. This condition had been brought about by the liberal terms which had been granted by the Wingfield-Nixon banks during the ballooning of Goldfield Consolidated, when the confederacy, according to common belief, was unloading millions of dollars' worth of stocks in the small companies and was using the proceeds to finance their purchase of the stock of several of the integrals that formed the big merger.

I began to get next to myself and to smell a rat. I had never had so much as an argument with either Mr. Wingfield or Mr. Nixon, had never been engaged in any business transactions with them, and the campaign against the trust company, which I felt sure had been conceived at the outset in the interests of the Republican political machine, I now suspected was part of a general scheme to get hold of anything and everything that was valuable in the camp. By smashing the Sullivan Trust Co. they could hurt the Democratic party of the state, with which we were affiliated, and for which it was currently believed we were supplying the sinews of war. By smashing us they might also cripple the bank with which we were doing business, and which in both Goldfield and Tonopah, particularly Tonopah, was a formidable competitor of the banking interests. And thus they might also facilitate a decline in the market which would shake out of their holdings borrowers at their banks.

I figured it out this way: Wingfield and Nixon knew that we had foolishly attempted to support the market for our stocks, that other promoters in Goldfield had done likewise, and that investors and brokers in Goldfield had borrowed heavily from all of the banks. John S. Cook & Co. were calling for more collateral from their customers, and real estate was being added to the pledges of mining securities. What more easy, even though diabolical, than to bear the market, shake out the stockholders in various important mines of the camp, take their stocks away from them by foreclosure, and get possession again, at bankrupt sale prices, of the millions of dollars' worth of securities which they had unloaded during the boom?

If this was the scheme of Wingfield and Nixon, what transpired could not have been patterned more perfectly.

Mr. Wingfield walked the streets day and night, armed to the teeth, and openly dared any of the miners to get him. He threatened another shut down, a reduction of wages, the installation of change rooms at the mines and other dire, inconvenient matters, all seemingly calculated to rouse the ire of the mine workers.

The miners fell for the bait, became belligerent and nasty and did things with which the community was not in sympathy. Day by day the situation became more critical.

During one of the shut downs which ensued, Senator Nixon revealed his hand by convening a meeting of the executive committees of the two Goldfield stock exchanges. He insisted that the exchanges close, arguing that the prices of stocks should be allowed to recede in sympathy with the labor troubles. No thought was his for the men of the camp who were committed to the long side of the market at boom prices and who had worked day and night to create the boom which had thrown into the laps of Wingfield and Nixon riches far beyond the dreams of avarice. The brokers refused to close the exchanges.

Goldfielders were slow to grasp the real import of what was transpiring. Things were very much unsettled. Optimism would rule today on apparently inspired rumors that the differences between the mine owners and the miners were about to be patched up. The next day gloom would pervade the camp because of the unfavorable action by the union on the peace plans. Nightly conferences were held. It was impossible to get an accurate line on the situation. Crowds gathered about Miners' Union Hall, where the meetings were held, and everyone sought something tangible on which to base his market operations. The officers of the union were in and out of the market, taking advantage of their official positions to anticipate every favorable or unfavorable development.

It was a critically sensitive market situation. The drift, however, was unmistakably downward. Values began to melt like snow in a spring thaw.

Through it all the Sullivan Trust Co. stood valiantly behind its securities in all markets where they were traded in—to the limit. I was bull-headed. I had never before been through a mining camp boom of such proportions, and I failed to recognize that a reaction must ensue, whether it was forced by Wingfield and Nixon or not. Tens of thousands of shares of Sullivan stocks were thrown at our brokers on the San Francisco Stock Exchange and New York Curb from day to day, and we took them all in, refusing to allow the market to yield to the pressure.

FROM CREDIT TO CASH

To convey an idea as to the standing of the L. M. Sullivan Trust Co. during this crucial period, I cite an instance. Logan & Bryan, members of the

New York Stock Exchange, Chicago Stock Exchange, Chicago Board of Trade, New Orleans Cotton Exchange and all other important exchanges, who conduct a leased-wire system from coast to coast at a cost of $300,000 per annum, and who have more than 100 correspondents in nearly as many cities, all of high standing as stock brokers, made a tentative offer to the Sullivan Trust Co. early in December to connect their wire system with our office in Goldfield and to give us the exclusive wire connection for Nevada at an annual rental of $100,000. This offer would not have been made if the credit of the Sullivan Trust Co. had not been maintained at high notch, or if I, personally, had not convinced men of substance that I was strictly on the level, past or no past.

Ben Bryan, the active member of this firm, was in Goldfield at the time. He asked as to our finances. There was present Cashier J. L. Lindsey of the State Bank & Trust Co.

"How much would your bank loan the Sullivan Trust Co. on its unindorsed paper and at a moment's notice?" I asked Mr. Lindsey.

"A quarter of a million or more," answered Mr. Lindsey.

This apparently satisfied Mr. Bryan.

Our rating in Bradstreet's and Dun's was "AA1." A private statement issued by Bradstreet was to the effect that while our rating was only $1,000,000 and we claimed a capital and surplus of only $1,000,000 at the time the rating was given, it was believed in Goldfield that we were worth much more, and that we had actually understated our resources because we considered it bad policy to divulge the great profits in the promotion business.

By December 15 the condition of the Sullivan Trust Co. had become about as follows: our $3,000,000 surplus had been reduced to $2,000,000 and all of this $2,000,000, plus the loss, was represented by all our own bought back stocks. We had no money, except about $50,000, remaining of the $300,000 borrowed from the State Bank & Trust Co. We were committed in excess of this $50,000 to brokers for stocks in transit, but by the crossing process we were able to maintain a chain that kept intact our reduced cash balance. We figured that a fresh loan of $300,000, additional to the $300,000 already obtained from the State Bank & Trust Co., would enable us to take up all of our paper and to discontinue the cross trades.

We promptly arranged for the loan, which Cashier Lindsey of the State Bank & Trust Co. informed us would be immediately credited to our account whenever we required the money. Interest charges were at the rate of one per cent a month in the camp at that time, and for that reason I did not ask that we be at once credited with the amount. I sent over to the State Bank

& Trust Co. another big batch of stocks, to be held as collateral against the promised loan, and got a receipt for it stating that it was accepted as collateral on our open loan account.

The market in Sullivan stocks had now steadied itself and it appeared that it would be impossible for any further selling of consequence to take place. We had bought back in the open market fully 50 per cent of all the stocks promoted by the trust company. Distribution of the stocks of our early promotions had originally taken place in such a broad way that it now appeared as if selling must necessarily become scattered. We felt somewhat crippled, but in no danger, and were still in the ring.

DOWN WITH THE SULLIVAN TRUST COMPANY

BY THIS TIME I was all in physically. I had a cyst of fifteen years' growth on the back of my head. It had become infected. I was threatened with blood poisoning. I suffered much pain. I had been on the desert for nearly three years without leaving it for a day. My associates insisted that I go to Los Angeles immediately for treatment and a rest. Believing that the trust company was secure, I made preparations to go. Before leaving I busied myself with the preparation of a dozen full page reading matter advertisements on Sullivan properties, which the Salt Lake *Tribune* and Salt Lake *Herald* had contracted to publish in their New Year's Day editions. These are an annual feature of those newspapers. I decided to make Salt Lake on my return trip from Los Angeles and be there on New Year's Day with our mailing list, to superintend the mailing of the papers to all stockholders in Sullivan properties. On account of the great value which we attached to the mailing list, I would not trust anybody but myself with the job. I spent Christmas in Los Angeles and arrived in Salt Lake on New Year's Day, ready for work.

I was busy in the Salt Lake *Herald* office next day when affable Peter Grant, a partner of Mr. Sullivan, with whom Mr. Sullivan had at the outset divided his interest in the Sullivan Trust Co., walked in. I asked Mr. Grant, who had remained at the helm with Mr. Sullivan while I was away from Goldfield, about business. He assured me that the loan from the State Bank & Trust Co. would not only be forthcoming, as needed, but that Cashier Lindsey had informed him that we could have $500,000 instead of $300,000 additional, if we actually had to have it, and that the bank would back us to the extent of a million in all, if necessary.

On calling next morning at the office of James A. Pollock & Co., our Salt Lake correspondents, I was astounded to learn that rumors had been telegraphed to them from San Francisco that our paper was being held up in Goldfield.

"That's nonsense," said Mr. Grant. "Why, Lindsey has given me his word, and there can't be a question about it."

"Maybe he has 'laid down' on us," I said, "and that would be ———!"

"Nonsense!" said Mr. Grant. "I'll telegraph him that in addition to honoring our Goldfield paper with the money we have borrowed from him, he must wire $150,000 to our credit in San Francisco, and you and I can jump on the train today and go to San Francisco and support the market right on the ground. If those rumors have spread around San Francisco a lot of short-selling will take place and the market will need support."

I agreed.

So confident were Mr. Grant, James A. Pollock & Company, and I that everything was right with us that we gave and they accepted a big supporting order to be used on the San Francisco Stock Exchange during the succeeding day while Mr. Grant and I should be on the train to the Coast city.

We arrived in San Francisco late at night. A number of brokers met us and conveyed the news that the State Bank & Trust Co. had "laid down" on us. In the meantime despatches to us from the cashier of the Sullivan Trust Co. had piled up at the hotel. He explained the situation, which was this:

All the trains carrying drafts in the mail to Goldfield had been stalled by snowstorms two days before New Year's. The next day was Sunday. Monday was New Year's Day, a legal holiday. Thus five days' mail had accumulated, and on Tuesday the delayed drafts were presented, all in a bunch.

L. M. Sullivan, president of the trust company, who was supposed to be on deck at Goldfield, was in Tonopah, where he was reported to be in imminent danger of arrest on the charge that during a New Year's brawl he had nearly brained a chauffeur with the butt end of a revolver. The bank people became alarmed.

In requisitioning the $300,000 we had stated that we would call for it piecemeal, as had been our custom in the past. The five days' mail had piled up drafts totaling nearly the entire amount. I was absent from Goldfield. Mr. Grant was away, and so was Mr. Sullivan. Employees were running the business. Cashier Lindsey concluded that we were overboard. On top of it all, Donald Mackenzie, the heaviest depositor of the State Bank & Trust Co., had that very morning drawn out a large sum, said to aggregate $400,000, and had it transferred to San Francisco. Our wires from Goldfield stated he had been frightened by rumors that the Sullivan Trust Co. was in trouble and that the State Bank & Trust Co. would be involved.

That settled it. The enterprise that I had built up from such a meager beginning into a $3,000,000 trust company crumbled in a heap and left us stranded on the financial shoals of an over boomed mining camp.

SOME HINDSIGHT THAT CAME TOO LATE

I ATTRIBUTE THE destruction of the Sullivan Trust Co. to six factors, namely, (1) politics; (2) blackmail; (3) lack of wide distribution of our later promotions, we having sold most of these stocks in large blocks during the exciting boom days through brokers to speculators instead of disposing of them in small lots direct to investors; (4) my lack of knowledge of markets and inexperience in market manipulation; (5) my own stubborn pride and optimism, and (6) the failure of the State Bank & Trust Co. to keep its pledge of assistance.

It is conceded in Nevada by all honest men that, without exception, all of the properties promoted by the L. M. Sullivan Trust Co. had merit, and that money was lavishly provided for mine development as long as the trust company was in existence. The properties were selected with great care. They were very much higher in quality than the average. Those at Manhattan are yielding treasure to this very day, and may make good yet in a handsome way from a mining standpoint. Those at Fairview bid fair to duplicate the performance. Had I kept out of politics, been a good market general, and taken cognizance of the fact that the law of supply and demand is as inexorable in mining stock markets as in every other line of human endeavor, I could have saved myself and associates from financial ruin.

It would have been the better part of valor to have emulated Bob Acres—back up and "live to fight another day." Instead, I attempted the impossible in my endeavor to stem the tide of liquidation, and exhausted our resources to the last dollar in buying back the Sullivan stocks at advanced figures over the promotion prices. I didn't know then, as I know now, that the accepted practice of the successful market operators is to go with the crowd—to help along an advance when the public is buying, and, with equal facility, to further a decline when everybody wants to sell. It was my first experience, and, like so many beginners, I was overconfident, lacking in judgment, and fatally ignorant of the finer points of the game.

The complete collapse of the financial structure I had labored so hard to construct came as an overwhelming blow to the camp and marked the beginning of the end of the great Goldfield mining stock craze.

Our enemies had overshot the mark. Public confidence was irreparably shattered by the smash of the trust company, and it would have been better for Goldfield and Nevada had Wingfield and Nixon possessed sufficient foresight to go to our rescue instead of facilitating our destruction. Money that had poured into the camp without cessation month after month for mine development started to flow the other way.

Less than a year later, when Wall Street's financial cataclysm put a quietus on market activities of every sort, the great fortunes of Wingfield and Nixon themselves hung in the balance, and had it not been for a quick transaction by which the United States Mint at San Francisco forwarded by express to Reno and Goldfield $500,000 in gold, the failure of Wingfield and Nixon and their chain of banks might have happened as a fitting climax to the scheme of aggrandizement which they had fostered.

It was rumored at the time that this money had either been obtained from the Government as a deposit for the Nixon National Bank in Reno or was obtained at great sacrifice from Wall Street bankers, and that only by virtue of Mr. Nixon's position as Chairman of the Committee on National Banks of the United States Senate was he able to get the Sub-Treasury in New York to instruct the Mint at San Francisco to supply the gold at this crucial period when fiat money was current in the East. Whether it was a government deposit or not, Senator Nixon got it—and he needed it.

Even to this day Wingfield and Nixon are engaged in an effort to shift the responsibility to me for the destruction of the great mining camp of Goldfield, which today marks the graveyard of a million blighted hopes.

On the eve of the Wall Street panic of 1907, every bank in Goldfield and Tonopah that had existed through the mining boom with the exception of those of Wingfield and Nixon, went to the wall, and every Goldfield broker, with one or two exceptions, went broke. The business interests of the camp suffered the same experience. Wingfield and Nixon succeeded in annexing the remnants of the Goldfield banking business, along with the control of nearly all of the Goldfield properties for which they had been seemingly gunning. Wingfield and Nixon are, in fact today in control of the political as well as the banking and precious metal mining industry of the state. They have triumphed, but Goldfield, except for the big mine and one or two others of little consequence which they do not own, has been throttled and is dying the death. Had Wingfield and Nixon played a broad gauged game, the camp would undoubtedly still be on the map and, instead of having only two or three mines, might now boast of thirty.

As quickly as possible I convened a meeting of the creditors of the Sullivan Trust Co., all of whom happened to be either Western brokers or banks. The market had gone to smash and our liabilities were $1,200,000. The assets, calculated at the low market price of the securities that was reached after the embarrassment was publicly announced, were still in excess of the liabilities. The creditors agreed in jig time that if we would turn over all of the securities they would accept 80 per cent of the net proceeds as full payment of our obligation and return the other 20 per cent to the trust company.

Thomas B. Rickey, president of the State Bank & Trust Co., was appointed manager of the pool, and was also elected president of the Sullivan Trust Co., which exists in moribund state to this day. Mr. Rickey had even a higher opinion of the value of the securities than we had, and he refused to sell any of them at the prices which then prevailed. He held on. During the bankers' panic of 1907 the State Bank & Trust Co. failed for about $3,000,000. The Sullivan mines were compelled to shut down. Mr. Rickey still held on. Manhattan, the mining camp, struck the toboggan. The boom in Goldfield securities collapsed at the same moment. The Sullivan stocks shriveled, like the rest of the list, to almost nothing.

As far as I can learn, neither the bank or broker-creditors nor any of the members of the Sullivan Trust Co. have ever received a dollar as a result of the settlement. Had the securities been disposed of immediately after the embarrassment, the trust company would have paid dollar for dollar. Those of the public who did not sell their holdings in the Sullivan companies when we were supporting the market to the extent of more than $3,000,000, lost most of their investment. Those who did sell—most of them—made money. The market value of these securities, at the height of the boom, was in excess of $5,000,000. The price paid for them by the public, as already stated, was in the neighborhood of $2,000,000.

After settling with the creditors of the Sullivan Trust Co. on the basis just outlined, I departed from Goldfield as broke as when I arrived there three years before. The only money I or my partners had drawn from the business during the life of the trust company was about $5,000, just sufficient to pay living expenses. My expenses to New York, where I went to have my head operated on—are you surprised?—were supplied by the proceeds of the sale of my seat on one of the Goldfield stock exchanges, from which I netted $400. I landed back in the big city with $200 in my pocket, the exact sum with which I had left town three years before.

My reward for three years of untiring work on the desert was a big fund of experience. Believe me, I thought it would hold me for a while! But it didn't.

6. Nipissing & Goldfield Consolidated

THE EMBARRASSMENT OF the L. M. Sullivan Trust Co., was disastrous to Goldfield. The decline and fall of the camp dated from that very hour.

The Goldfield *News,* of nationwide circulation in those days and up to then unshackled, sought to stem the tide. It published a double-leaded editorial, in full face type, setting forth that the Sullivan Trust Co. had gone down with its flag nailed to the masthead of a declining market and had lost its last dollar supporting its own stocks.

The camp took courage. Soon it became evident that the initial smash in stock market values was not sufficient to convince the natives that the death knell of the market for its long line of mining securities had been sounded.

The population of Goldfield was 15,000. Its life could not be snuffed out in a day. Great was the depreciation in the market price of Goldfield mining issues, but not to an extent as yet that indicated the almost complete annihilation of values which followed. Final destruction for the general list, with some scattering exceptions, came only after a starving-out siege on the part of investors, who refused to commit themselves farther and gradually resorted to liquidation.

Listed Goldfield securities, nearly 200 in number, and valued in the markets at above $150,000,000 during the boom, had within two months shown a falling off of $60,000,000 in market value, but the list on the average was still quoted higher than the promotion prices.

On January 18, 1907, fifteen days after the newspapers throughout the land carried front page stories of the failure of the Sullivan Trust Co., the stocks promoted by the trust company were still in demand in all mining share markets of the country at an average price not below that at which original subscriptions were accepted from the public.

Jumping Jack, promoted at 25 cents, was quoted at 30 cents bid. Stray

Dog Manhattan, promoted at 45 cents, was in demand at 49 cents. Lou Dillon, promoted at 25 cents, was still wanted at 26. Indian Camp, sold originally to the public at 25, was quoted at 85 bid. Silver Pick Extension, promoted at 25, was 21 bid, a loss of four cents from the promotion price. Eagle's Nest Fairview was quoted at 25, off 10 cents from the promotion figure. These prices represented terrific losses from the highs that had been reached during the height of the Goldfield boom, yet the average market price was still above the subscription price of the shares at which the public was first allowed to participate. A remarkable part of this demonstration was that for twenty days no inside support had been lent to these stocks. The Sullivan Trust Co. being in trouble, the markets had been left to the mercy of short sellers and market sharp shooters generally.

Having settled the trust company's liabilities of $1,200,000 by tying up in trust all of its securities and the other assets, of which the creditors agreed to accept in full quittance 80 per cent of the proceeds and to turn back to the trust company 20 per cent, I returned to New York during the last week in January. I was again out of a job—and broke.

I visited the officers of mining stock brokers in Wall Street and Broad Street. Wherever I went a hearty handclasp was extended. Not one of the Eastern stock brokers was involved to the extent of a single dollar in the Sullivan Trust Co. failure.

The brokers were convinced that the embarrassment was honest. The trust company's credit had always been good. Had the failure been meditated, I could have involved Eastern brokers for at least $1,000,000. Because I didn't, New York brokers were not slow to express their good feeling. A number of them offered to extend a helping hand did I wish to embark on a new enterprise.

Peculiarly enough—or shall I say, naturally—after tossing off the trust company's millions, of which half were mine, in a vain endeavor to support the market for its stocks, I was as full of spirit as the month of May. I had been broke before, and the sensation was not new to me. Withal, I had profited. A new fund of experience was mine. Even though I had not gathered shekels as a result of my hard work in Goldfield, I had learned something —I had acquired the rudiments of a great business.

Goldfield had been the mining emporium—the security factory. New York was the recognized market center. Market handling had been my weak spot. I now had a chance to witness the performance of some past masters in the art of market manipulation, and I tried to make the best of the opportunity.

I watched intently the daily sessions of the New York Curb. I was in and out of brokerage offices hourly. Nothing that transpired escaped me.

Within a month I heard enough and saw enough to convince me that, daring as were the operations of the mergerers and waterers of Goldfield Consolidated, in that they ballooned the price of their security at its inception some $29,000,000 (400 per cent) above the accepted intrinsic worth and were able to get the public in at top prices, their activities were but amateurish when compared with the stock market campaign in Nipissing, which was now transpiring on the New York Curb.

In the Nipissing campaign tens of millions of the public's money went glimmering, several great promoters' fortunes were reared as by magic, some big names and big reputations were tarnished, and dollars in $1,000,000 blocks were juggled like glass balls under the touch of sleight of hand performers.

AN ORGY IN MARKET MANIPULATION

THIS MARKET MELODRAMA was well staged. It had a sensational start off, and action was at high tension every minute. The performance had covered a period of seven months when I arrived in New York, and was reaching its climax. It was a wild orgy in market manipulation and money fleecing that had no parallel in history from the early Comstock days up to and including Greenwater. As a mining stock boom it was a dizzy, bewildering success— full of red fire and explosions to the last curtain climax.

W. B. Thompson, Montana mine promoter and money-getter; Captain Joseph R. De Lamar, famed as a daring adventurer on land and sea, and recently a highly successful financier, mine owner, stock market operator and art collector; John Hays Hammond, mining engineer, promoter, politician and ambitious society leader; A. Chester Beatty, millionaire mining engineer, and the seven Guggenheim brothers, were in the all star cast. Mr. Thompson, by reason of the fact that he was market manager, was most under the spotlight, although at times he was obscured by the others.

Mr. Thompson was a product of Butte, Montana. Early in the game he had learned the Wall Street lesson that stocks are made to sell. Born and reared in Butte without the aid of a silver spoon, he had never been in the money before coming East. The great paystreak in the East apparently looked better to him than the paystreaks that some of his Butte neighbors had missed in their deep mine operations. He was an ideal man for the Nipissing job, as subsequent events in his career thoroughly confirm. Of a school that believes money in hand to be worth more than certificates in the box, Mr. Thompson's route from Montana to Broad Street was via Boston, where he made his first stake by marketing stock in the Shannon group of mines.

Nipissing & Goldfield Consolidated

When the Cobalt excitement was in its infancy Mr. Thompson took a run up to the camp. The Nipissing mine was about the best thing in sight. It was producing real silver. The company was owned by a little club consisting of E. P. Earle, specialist in rare metals, Captain De Lamar, millionaire soldier of fortune, E. C. Converse, banker and steel magnate, Ambrose Monnell, R. M. Thompson, Joseph Wharton, since deceased, of Philadelphia, and Duncan Coulson, a rich Canadian lawyer. Considerable silver was being produced. The veins, however, were exceedingly narrow, not more than a few inches wide. It was impossible to block out ore to an extent that would warrant any opinion as to the real measure of the mine's riches. The gentlemen owners were not averse to giving Mr. Thompson an option on 100,000 shares of treasury stock of the 1,200,000 five dollar shares ($6,000,000), at $2 a share, when he made the proposition, and another 100,000 shares at $2.50. Later, they sold him a call on 50,000 or 100,000 shares around $7. All of this happened in the summer of 1906, six months before I reached New York and at a time when the country was giving indication of going mining stock crazy, Nevada stocks having advanced on the New York Curb in the Goldfield boom hundreds per cent.

After the Goldfield boom had gained terrific headway, during the fall of 1906, when Mohawk was climbing from 10 cents per share toward the $20 mark, which it reached during the climax, the Cobalt mining stock excitement spread like wildfire. A sudden demand sprang up for Nipissing shares. Mr. Thompson, about this time, connected himself with the old established and conservative banking house of C. Shumacher & Co. on Wall Street. The affiliation was calculated to give the promoter of Nipissing stock much standing. The move served well its purpose. The public grabbed at the shares. The price jumped to $4.50 in a jiffy. Mr. Thompson began to let go of stock after the $4 point was reached. He was making a killing, but fed out his optional stock very cautiously at the rate of about 5,000 shares daily, each day at an advance. By the time the price reached $7 Mr. Thompson got suspicious. There was something about the play he could not understand. He had not found it necessary to do much laundry work on the Curb market. Every time he offered stock it was lapped up silently and completely. Every time his brokers opened their mouths to sell the certificates they were gobbled.

Mr. Thompson stopped putting out any more stock and streaked it up to Cobalt to see what was going on. He had a hard time laying hold of the inside facts, but learned enough to satisfy himself that rich ore had been encountered at depth. He discovered on his return to New York that Captain De Lamar had been buying that cheap stock through S. H. P. Pell & Co. and was even then the heaviest individual holder, a position contested only once during the whole campaign, and that by a rank outsider operating through Eugene Meyer, Jr., whose name has never been publicly mentioned as having anything to do with the gamble. This unknown was a quiet, mild spoken, college-bred gentleman. He pulled down $1,500,000 in Nipissing—and kept it.

Upon the return of Mr. Thompson from Cobalt the promoters warmed up to their job. The manipulation which had been begun in a comparatively modest way now showed the spirit of the gambler who plays the ceiling for the limit. New market boosting accessories were called into use. They did their work. The game waxed hotter and hotter.

THE GUGGENHEIMS ENTER NIPISSING

Boom! boom! boom! went Nipissing. By the time the price crossed $20 the gamblers and speculators of two continents were on fire with excitement. Presently it became noised about that the Guggenheim family had taken an option on 400,000 shares of Nipissing stock at $25 a share, making the investment $10,000,000, and putting a valuation of $32,000,000 on the property. Furthermore, it was announced that the deal had been made on the report and advice of John Hays Hammond, the international mining engineer, crony of Cecil Rhodes and famed as the head of the profession. As a part and parcel of the remarkable story, it was authoritatively stated that the Guggenheims had paid $2,500,000 cash for the option. W. B. Thompson was said to have negotiated the transaction.

Confirmation of the deal set the gamblers crazy. There could be no risk in following such leadership as the Guggenheims', endorsed by the eminent Hammond. The market boiled up to $30 and then majestically boomed to $33.25. Transactions in this single issue totaled hundreds of thousands of shares a day. Waiters, barkeepers, tailors, seamstresses and tenderloin beauties competed with bankers, merchants, professionals on the regular exchanges, and even ministers of the Gospel, for the privilege of buying Nipissing shares on a valuation of more than $40,000,000 for the mine.

On the way up the original bunch of insiders floated out of their holdings. Most of them had cashed in under $20. Some of them stayed out; others went back, and, like the moth, got burned. W. B. Thompson, it is said, parted with the bulk of his 250,000 to 300,000 shares at from $24.50 up, cleaning up for personal account between $4,500,000 and $5,000,000, according to the estimates of close friends then in his confidence. Never was there a cleaner case of finding money for Mr. Thompson. The manipulative campaign, of which he was made manager, was a giant success. The only ability or skill needed, after the Guggenheim deal was made—brilliant deal from a market standpoint!—was the sense to hold on to his optioned stock until his associates, the Guggenheim following, and the public made a rich, ripe and juicy market for it.

Mr. Thompson subsequently participated in Cumberland-Ely, El Reyo, Inspiration, La Rose, Utah Copper, Mason Valley, and other mining promotions, and is now rated at $10,000,000 to $12,000,000. He is generally prominent at the nutritious or selling end when a good market exists and is now head of a New York Stock Exchange brokerage and mining promotion firm which publishes its own newspaper.

But what happened to Nipissing? Plenty, and then some, happened. As noted, the stock mounted by flying leaps to $33.25, stayed well above $30 for quite a while, and began slowly to recede. Complacent in the consciousness that they had the biggest silver mine in the world, the Guggenheims allowed all of their friends to share in their good fortune.

Of a sudden, stock from mysterious sources began to press on the market. It came in great quantity and without let-up. Suspicion was aroused in the Guggenheim camp. They despatched A. Chester Beatty, one of their very best expert engineers, and a former protégé of John Hays Hammond, to Cobalt to smell out the trouble. The text of his report was never printed. It didn't have to be. The facts beat it in.

Much of the showy mineral, on which glowing reports as to the fabulous value of the property had been based, contained little or no silver. It was *smaltite*, an ore of the metal cobalt, closely resembling many of the silver ores.

The story was given out that Mr. Beatty reported adversely on account of the unfavorable showing made by mine developments carried out subsequent to Mr. Hammond's report. The miners had run into non-productive calcite a few hundred feet down, it was said.

As a matter of fact, because of the limited amount of all underground development in the interim, there could have been no condition observable in the property as a whole when Mr. Beatty made his examination that was not equally apparent when Mr. Hammond made his report.

The talent jumped to the conclusion that the mine was a "deader."

Many millions in silver bullion have been taken from the property since then, and it is still a great producer, but this is another and more prosaic story. This deals with the stock gambling feature of the record.

Scenes of the wildest disorder were witnessed on the Curb in those days of 1907 soon after my return from Goldfield. The Guggenheims laid down on their option, getting out as best they could. According to published reports, they charged to profit and loss the $2,500,000 originally put up, besides paying the $1,500,000 to $2,000,000 in losses of personal friends for whose misfortune they felt personally responsible. Be that as it may, the Guggenheims emerged from the campaign with damage to their market reputation and standing from which they have never fully recovered. Previous to their acquaintance with the Cobalt bonanza, they had a blindly idolatrous following that would have invested hundreds of millions on a tip from them. They have never regained the position in this respect they then held.

NIPISSING ON THE TOBOGGAN

THE PRICE OF Nipissing tobogganed from $33 to under $6 with terrific speed. W. B. Thompson and his associates, who had unloaded their holdings on the way up, were reported to have taken advantage of the Beatty report and to have sold the market short on the way down, making another clean-up of millions. The stock hit a few hard spots on the descent, but when the wreckage was cleared away and the dead and wounded assembled, there wasn't hospital or morgue space to accommodate half of them. The final carnage and mutilation was shocking beyond description.

The public had once more been landed with the goods. It had eaten up Nipissing stock on a $43,000,000 valuation which broke at $7,000,000 or $8,000,000 within the space of a few days. This $35,000,000 slaughter represents only a fraction of the actual losses, for fabulous amounts were sacrificed in marginal accounts. The daily aggregate of open accounts in Nipissing during the months of keenest excitement probably averaged not less than five times the total capitalization. Actual losses were therefore far larger than would appear from a merely superficial calculation. The public contributed $75,000,000 to $100,000,000 to its Nipissing experience fund.

There has always been more or less mystery as to just what John Hays Hammond said orally to the Guggenheims to lead them into the crowning humiliation of their business career. It did not appear in his written and published report, for in that document is to be found a neat little hedge to the effect that *if* conditions as revealed to him were maintained, the values would be, etc., etc. That little "if" was the Hammond saving clause, although it did not save that $1,000,000 a year job of his, about which some of his admirers have liked to talk in joyous chorus, nor did it save the public from massacre.

Another Nipissing mystery is the sustained professional and personal cordiality still existing between the eminent John Hays Hammond and the scarcely less eminent A. Chester Beatty. For a little while after Mr. Beatty had to turn down his chief their relations appeared to have been strained. But this was not for long. Mr. Beatty also severed connections with the Guggenheim payroll, and the two great engineers were soon again, and are now, on the best of terms.

On rainy days when the tickers drone along and there is no exciting news, evil-minded derelicts of the memorable Nipissing campaign are prone to figure how much a man might have made in the market with a foreknowledge of the two adverse reports and to figure on the sporting chances for a double cross that such a situation would hold.

Scandal mongers, too, who have watched closely the friendship which exists between W. B. Thompson and John Hays Hammond often ask unkindly what has cemented the bond between the two. Recently, when the Rocky Mountain Club needed a new clubhouse, Messrs. Hammond and Thompson subscribed an equal amount—a goodly sum it was—to build it.

They are seen much together in public and seem to have many tastes in common. Mr. Thompson, whose strangely fortunate campaign in Nipissing on the New York Curb was helped to a triumphant promotion climax by the Hammond report to the Guggenheims, bears Mr. Hammond no ill-will for that—and who would blame him for the kindly feeling?

WHO GOT THE $75,000,000

BUT WHAT OF the public? It played $75,000,000 to $100,000,000 into the game, and has never yet learned who got it. Who did get it? Some of the details of the grand separation scheme have been set forth in the foregoing, but nothing like enough to satisfy the curiosity of the public who footed the bill, paid the freight, contributed sucker-toll for the whole prodigious sum.

Did the author of the report on the strength of which tens of millions were plunged on Nipissing by an army of deluded investors and speculators ever suffer in fortune by the mischance or misshot, or whatever name you may give the come-on document? Not that you could notice. True, he gave up his alleged $1,000,000 job with the Guggenheims. But is he not a heavy contributor to the Republican national campaign fund, a close personal friend of the Administration, and did he not represent this great government as Special Ambassador at the Coronation of England's King? Was he not talked of as running mate for Mr. Taft, and did he not organize the National League of Republican Clubs two years ago? He is tremendously rich and round-shouldered under a mountain high burden of honors.

Every mother's son of the old Nipissing crowd is at this very hour up and at it in regions where the public's money flows. Many of them still have

GOLDFIELD DAILY TR[IBUNE]

METAL MARKETS
NEW YORK, November 25
—Silver 55⅜ cents; copper
quiet, $13.50 to $13.30; lead
firm, $4.40 to $4.50

TUESDAY EVENING, NOVEMBER 28, 1911

VOL. VI No. 56

BOOSTING STORIES WRITTEN BY RICE AND HIS FOLLOWERS TO HORNSWOGGLE THE PUBLIC

Market Letters Pretended That Rice, Scheftels et al. Had No Interest Except as Brokers In the Stock of Rawhide Coalition Which Kited From Fifteen Cents to $1.50.

Special to the Tribune.

NEW YORK, Nov. 28 — At yesterday's hearing in the trial of Rice, Scheftels and their associates, the government introduced a bundle of market letters sent out by the Nat C Goodwin company, booming Rawhide. The government also presented copies of the advertisements and the "boosting" stories printed in the Mining Financial News of which Rice was editor, exploiting Rawhide Coalition as the best mining proposition on the market. The district attorney pointed out statements in the market letters intended to convey the impression that the defendants had no interest in this stock except as brokers, and that they were entirely disinterested in the extrava...

...over National bank in New York for Rice.

The defendants' counsel asked Scheeline if the Rawhide mine was of any value. The district attorney objected to this question unless Scheeline could first qualify as a mining expert. He said the government would have recognized experts pass on this subject in due time. The defense claimed the right to question Scheeline on this point, since the government had made the charge that the defendants had unloaded Rawhide stock on its customers. The government replied that mining experts were the ones to pass upon the value of the property and not bankers.

DRIV[EN] OF [FR]OM STATION

SUED FOR LIBEL

The Denver Mining Record Hailed to Court by the Goldfield-Tonopah Advertising Agency.

Christopher F. Clay, the noted Denver lawyer, has been instructed by telegraph by the Goldfield-Tonopah Advertising Agency, to bring suit at once against the Denver Daily Mining Record for libel, says the Goldfield Sun of last Monday. The publication of a bitter attack that recently appeared in its columns, and of which the Record has sent marked copies broadcast, is the cause. A. G. Denniston, president of the agency, said today:

"We will prove that the Denver Mining Record is engaged in a dastardly and diabolical attempt to destroy our good name because the agency refused to recommend the Record's columns as a good advertising medium to reputable people, for two reasons, one of which was that the sheet has the reputation of being a 'wildcat' organ, and the other that it refused to conduct its business with us in an honorable way. We have a dozen letters from it opportuning us to send it business, as late as three months after we discontinued all business relations with it, and we will quickly prove the statement that it turned down our business or found fault with our news service, to be a lie. The Denver Mining Record has long been a stench in the nostrils of Goldfield because of its catering to fake mining companies, and we cut it off our list, and not it us.

"The statements made by the Denver Mining Record attacking the integrity of our Mr. G. G. Rice, are vile calumnies. I have known Mr. Rice for fifteen years, and I know that at least nine statements out of ten that they made regarding him can be disproved in court in five minutes.

"A crab is trying to bite an elephant's toe, but in this court proceeding we shall treat the Record as an elephant, and if the newspaper ain't already mortgaged above its value, we shall have a good chance of owning it in the near future. We guarantee to give the Record and the pair of buccaneers that we learn are behind the attack as big a run for their effort as they ever got in their lives."

Asked who the two buccaneers were, Mr. Denniston said:

"One used to live in Oregon. He withheld from our agency $500 right fully belonging to us. The other comes from Los Angeles, and he is now engaged through the Mining Record's columns in foisting on the public a raw prospect as a mine. He is selling shares at 25 cents, or a valuation of $250,000 for the property, and you couldn't get a bid of $10,000 on it to save your life if you tried to sell it to a mining man today. The postal authorities have had him under their surveillance for weeks. He was told to stay away from our office a fortnight ago. Fearing an expose of their methods through our news bureau, these two seek with the Denver Mining Record to undermine us."

Accurate Information.

If you wish to know the actual prices and to see actual transactions in all the Tonopah, Goldfield and Bullfrog stocks, write to D. G. Doubleday, banker and broker, 329 Pine street, San Francisco, for his market sheet showing actual transactions and actual prices. You can tell by his market sheet exactly the difference between the stocks which are active and those which are inactive. If you hold any Tonopah, Goldfield or Bullfrog stocks, you should have his sheet showing the market.

History of the Desert.

A free booklet, describing the discovery and development of Tonopah, Goldfield, Kawich and Bullfrog, is being distributed by Frank L. Kreider & Bro., brokers, Merchants Exchange, San Francisco. A copy of this booklet should be secured by every investor in these stocks of the districts named, as it contains valuable information concerning the leading properties.

J5·20

Subscribe for THE MINER.

AUGUST 3, 1918

RICE ARRESTED ON FRAUD CHARGE

FORMER GOLDFIELD PROMOTER, THREE TIMES CONVICTED, AGAIN DETAINED

NEW YORK, July 31. — A list containing the names of 102,000 persons whom Jacob Simon Herzig, who calls himself "George Graham Rice," apparently regarded as prospective purchasers of stocks has been seized by federal officials who arrested him Monday night on a charge of using the mails to commit fraud, the authorities announced. The persons mentioned live in various parts of the country.

Herzig, who has been involved, officials state, in other promotion enterprises which resulted in his conviction for criminal offenses, is held on a charge of accepting money for stock of the American Car & Foundry company, which he did not deliver to the purchaser. The federal grand jury yesterday began consideration of his case.

George Graham Rice constantly made the headlines while promoting Nevada and Death Valley mines

GEO. GRAHAM RICE.

A SUGGESTION FOR GRAHAM RICE

Herzig from Sing Sing, alias Rice from Reno, breaks the news through his market letter, the Nevada Mining News, that he is on the verge of blotting upon his clients a series of articles concerning men he has known in Nevada. The strange thing is that this parvenu imposter, this egotistic upstart, who classes himself with Mark Twain, does not exploit his own record. Modesty may forbid an autobiography, therefore the task might be subject. We suggest for author the editor of the Death Valley Chuckwalla, whose knowledge of the subject is displayed in the following article from the current number of that magazine, under the caption,

TWO DESERT JACKALS.

"Not so very many weeks ago two jackals at Goldfield who have been trailing the golden lion of this desert for several years had grown suddenly so felt and fat from the pelf of their plunder that they became indiscreet in their chase, a bit more unscrupulous in their manner of hunting and quite obnoxious to the decent people in the town. Their lair in the camp head, we do not care to repeat the odious detail of this preamble. But had been disturbed by a young writer from New York, who told an unvarnished tale of the ways of these prowling scavengers of the mines. He showed up the past records of the members of this gang and the savor of this story was not unwelcome to Goldfield. In a day the sheep's garb was torn from these thieves of the night and each individual member was compelled to hit the trail for darker regions. All this was passed into the history of Goldfield and the desert as the expose of the L. M. Sullivan Trust company.

"Because the above is history and not news, because the fiasco of the Sullivan Trust company gained early last winter much publicity, and because the said trust company was reorganized with honest men at the head, we do not care to repeat the odious detail of this preamble. But because these same crooks are again at their work in a different town, under different guise and are again plying their jackal methods with new and clever variations we wish to say something about it them. Possibly when L. M. (Shanghai) Sullivan of notorious fame in Portland, and his brainy ex-convict pal, George Graham Rice, were kicked out of the Sullivan Trust company of Goldfield by their creditors, many who had followed the expose of this fiasco wondered where next these oily, slimous gentlemen would begin operations. In what particular spot in this country they would materialize with new schemes of obtaining the gullible public's easy money. While it was very apparent... was only a heavy... weight type of Bill Sykes... able desire to follow the sh... Mr. Rice would soon evolve... the ring as ardent pickers... the curious on this point... curiosity satisfied. Also w... to our friends who may pe...

"A few days ago the... in the form of a minin... News. Its form was th... editor Merrill A. Teagu... on the editorial page b... editor stated in his co... the Nevada News Bur... rowth from the old she... L. M. Sullivan and... bureau. It was the... the halcyon days of... reports about the si... cated. This new s... plied most of the Ir... ivan stocks were a... tion. The appea... Nevada publi... the columns... d still anoth... intrinsic... tal is an e... public... eas of... d Sull... was u... of the... peared in... a correctness... by these proceedings?..."

NEW YORK PAPER AGAIN UNMASKS EX-CONVICT RICE

A dispatch from New York, under date of November 9th, says:

"Nat Goodwin and his partner, Jacob Herzog, alias George Graham Rice, who is 'No. 4018' on the Elmira Reformatory records and 'B-1514' at Sing Sing, are arraigned in an issue of the Mining and Engineering Journal for their method in Wall Street. According to the mining journal, Rice and his associates obtained options upon an enormous number of shares of the Ely Central Copper Co., believed to be more than 750,000, at or about 50 cents a share, and are selling it to the public at a greatly enhanced price by an extensive advertising campaign. The quotation today was $4 a share. The Mining Journal declares the property is practically worthless."

"Rice himself is of such evil r... tation that he cannot appear o... in any business that he is conde... and consequently requires a... head like B. H. Scheftels... made his debut in New Yo... nection with the operati... hide' Coalition, one of F... affairs, which was a f... unsavory feature of th... last winter. After th... Central business wa...

"Along with Rice a... Nat Goodwin, the actor,... on the letterhead of B. H. Sc... & Co. as vice-president of the corporation. Previously Goodwin was associated with Rice in the Nat C. Goodwin Brokerage Co. of Reno, Nev..."

lished a brokerage business in Chicago and fell in with Rice by marketing stock of some of the Nevada wildcats of the latter. He removed to New York early in 1906.

THE TONOPAH MINER, SATURDAY, APRIL 10.

Cited For Contempt.

Judge Langan of the First Judicial District Court Ormsby County has cited George Graham Rice and Merr A. Teague, editors of Mining Financial News, published a Reno, to appear before him on April 24th and show cause why they should not be punished for contempt of court. The alleged contempt is probably based on an open letter, in which the methods employed in handling the affairs of the busted banks are severely criticised. It is really too bad that any paper should fall so low as to be guilty of speaking a word in support of the unfortunate depositors who have been deprived of their money, and that it should so far forget itself as to dare criticise any act of a judge, bank commission or bank examiner. We are inclined to think that the wayward editors mentioned are perfectly willing to be cited, for the article published was a strong one, and it is not probable that it would have appeared in their paper unless they had ample proof of its correctness. Who knows but that some light may be shed by these proceedings? We may be able to learn why such a determined effort is being made to oust the present receivers and replace them by former officials of the defunct institutions. It will be remembered that State Bank Examiner Hofer ordered C. H. Phillips, an official of the bank, to take charge of the busted bank had been appointed. Then again, it is a matter of record that Judge Langan appointed George Hall, another official of the busted bank, to act as receiver. Can it be possible that all the talent in the State was previously employed by the ruined institutions? Let the good work go on. If we see in the hazy distance a pandora box a grip on the property. It was a good old cow to milk. E. P. Earle, who was president of Nipissing in 1906, headed the company four years later. Captain De Lamar slipped down and away (he's now in on the extravagantly touted Porcupine Dome Mines Co.), and so did E. C. Converse, whose time is all taken up managing the Stock Exchange banknote engraving monopoly and a couple of banks and trust companies. W. B. Thompson, who came into the Nipissing directory in 1907, still sticks in spite of the awful experience of 1906-07.

Has an outraged government ever raised hue and cry against these eminent captains of industry? Not yet, nor soon.

What difference is there between the respectable multi-millionaire bankers putting across a losing promotion and the little fellow? Both may be equally honest or equally crooked, yet in equity both are entitled to the same treatment and the same consideration. Their operations differ only in degree. The aim of each is to get the public's money. And the big fellow is more dangerous by a hundred thousand degrees.

Where does real tangible evidence of a conspiracy to defraud in Nipissing exist? Does *any* exist? Now I venture to say that you could put on the scent any young man who is a graduate of the public schools, and within thirty days he would obtain evidence to prove to any jury in the land that the manipulators of that stock used improper measures to get the public's money.

A scrutiny of the files of the newspapers during the progress of the malodorous Nipissing campaign reveals many strange happenings. It shows, among other things, most remarkable willingness on the part of financial writers for the press of that day to say every possible good word for the manipulators and to feed the public appetite for sensational gossip concerning the gamble.

How this was done is easily understood by those familiar with Wall Street publicity. It was an open secret on the Street at that time that many writers for the press were subjected to strongest temptations to lend their hand to the game of publicity. The columns of the daily newspapers carry in themselves evidence to show that the attempts were not always in vain.

One little story will illustrate the methods employed. The business manager of a widely known and reputable daily financial publication was stopped one day by a man active in Nipissing and told he had been put into 500 shares of the Nipissing stock at the market price when the stock was still selling under $10 and at the time when it was being groomed for the terrific rise which followed and which did not culminate until $33 had been passed. The newspaper man was not above making a turn in the Street, but he objected to taking it that way. He politely turned down the proposition, saying that he did not wish any part of it.

The tempter then went to him on another tack, agreeing to carry the stock for him, so that he would have no risk whatever, at the same time remarking that, in turn for the favor, generous recognition in the news columns of the publication, in support of the Curb campaign, would be expected. Again the newspaper man declined, this time with unmistakable emphasis. He intimated cannily that while he might be taken on he might not be told when to get off, adding that he might be discharged if he fell for anything of that sort.

When the market price toppled from $33 back to around $6 this man's newspaper did not carry any frontpage story denouncing the outrage upon the public.

I do not know that the manipulators of Nipissing got to his employers, but I do know of some newspapers in New York which pose before the public as embodying the very highest type of newspaper morality and which have at their head, either as part owners or as editors, men who were taken in hand by Wall Street magnates at a period when they were dependent for their daily livelihood on their weekly wage, and were lifted into the millionaire division by being put into good things. Do you suppose newspapers presided over by those men are going to say a word against the enterprises of their benefactors? Conversely, if their benefactors happen to be bothered by any man whose business purposes run contrary to theirs, how far, do you think, these gentlemen of the press would go in their own news columns to poison the public mind against the enterprise of their patron's enemy?

When I witnessed the climax of W. B. Thompson's marvelously successful campaign in Nipissing on the New York Curb, I was fresh from Goldfield. My recollection is that my chief thought at that time, with the Goldfield Consolidated swindle fresh in my mind, was simply that the Western multimillionaire highbinder promoter didn't class with his Eastern prototype. Indeed, the two appeared to be of different species, as different as the humble but noisy coyote from the Abyssinian man-eating tiger.

The late spring of 1907 found me back in Nevada. I selected Reno as a central point for residence and decided to locate there. Eastern stock markets appeared to be beyond my ken. It seemed quite apparent that the Western game, as compared with the Eastern, was one of marbles as against millions. In New York's financial mart I felt like a minnow in a sea of bass. Without millions for capital, Nevada appealed to me as a more likely field of usefulness. I believed in Nevada's mineral resources. Having seen Goldfield evolve from a tented station on the desert with a hundred people into a city of 15,000 inhabitants; from a district with a few gold prospects into a series of mines producing the yellow metal at the rate of nearly $1,000,000 a month, I was enthused with the idea that there were other goldfields yet unexplored in the battle born state and that opportunity was bound to come to me if I pitched my tent on the ground.

THE WONDER MINING CAMP STAMPEDE

I WAS BACK in Nevada just a week when a stampede into a new mining camp called Wonder took place. I was quick to join in the rush. The Philadelphia crowd who owned control of the big Tonopah mine had annexed a property there which they named the Nevada Wonder. It boasted of a big tonnage of a low-grade silver and gold ore.

On arrival at Wonder I found my former Goldfield partner, L. M. Sullivan, on the ground. He entreated me to allow him to cut in on any deal I made. A bargin was struck. He agreed to advance all the money and I was to receive half of the profits for my work. The corporation of Sullivan & Rice was formed. We purchased the Rich Gulch group of claims, a likely piece of ground with a well defined ledge, and incorporated the Rich Gulch Wonder Mining Co. A company with the usual million share capitalization was formed to operate the property. A high class directorate was secured. T. F. Dunnaway, vice-president and general manager of the Nevada, California & Oregon Railroad, accepted the presidency. Hon. John Sparks, Governor of Nevada, became first vice-president. U. S. Webb, Attorney General of California, accepted the second vice-presidency. D. B. Boyd, for twenty five years successively Treasurer of Washoe County, Nevada, was made treasurer.

The first advertised offering of treasury stock of the Rich Gulch Wonder carried the names of forty leading mining stock brokers, situated in various cities stretching from New York to Honolulu, who had signified over their signatures their willingness to undertake the sale of treasury stock at 25 cents per share on a basis of 20 per cent commission. The first thousand shares of treasury stock at 25 cents was sold to Superintendent McDaniel of the Nevada Wonder mine. This convinced us that we had a good prospect.

I had my doubts about the succesful promotion of any Nevada mining company at this period, because of the terrific slump which was transpiring in Goldfield issues and also because of the smack in the face that mining stock investors had just received in Nipissing. It was my idea that if the Rich Gulch Wonder made any money for us the cashing would have to be delayed until mills were erected and the property became a producer. I was willing to go ahead on that basis.

The sale of treasury stock was slow, but sufficient was disposed of to warrant the expense for mine development of at least $2,000 a month for six months, and that appeared far enough to provide for in advance.

Pending the making good of this proposition in a financial way, I determined I would help finance a newspaper publication at Reno which would give to mining stock speculators an unbiased statement of mining and market conditions as they existed. In the mining camps it was considered tantamount to financial suicide for the home publication to reflect on the merits of any locally owned property. Strictures were looked upon as knocks, and knockers are taboo in mining camps. Moreover, mining camp papers could hardly make both ends meet at the time without support from inside interests, and unprejudiced statements of fact that were detrimental to a local property could hardly be expected.

Merrill A. Teague was made editor of the new publication, which was called the *Nevada Mining News*. Mr. Teague had just blown into Reno from Goldfield where he had been connected with the Nevada Mines News Bureau, a daily market sheet. Before coming to Nevada he had served in an editorial capacity on the Baltimore *American* and the Philadelphia *North American*. Mr. Teague is the possessor of a facile pen. At $50 a week, which was his stipend at the beginning, I was convinced that the *Nevada Mining News* had a cheap editor. When news was scarce he could write more about nothing than any man I ever met before. Incidentally, he could go further without finding a stopping place in a crusade than any man I had ever bumped up against. That was his drawback. However, compared with the work of other newspaper men then employed in Nevada, his stuff was in a class by itself and was commercially very valuable.

TEAGUE ATTACKS SENATOR NIXON

Mr. Teague was on the job just a week when he cut loose with an attack on United States Senator George S. Nixon of Nevada in a front page story headed "Goldfield in the Grasp of Wall Street Sharks." The article declared that Senator Nixon, needing $1,000,000 to conclude the merger plans of the Goldfield Consolidated, had got it through B. M. (Berney) Baruch of the New York Stock Exchange, factotum of Thomas F. Ryan, at terrible cost. The loan was made at a time when Goldfield Consolidated was selling around $10 per share. In consideration for the loan, Senator Nixon, acting for the company, gave Mr. Baruch an option on 1,000,000 shares of treasury stock of the Goldfield Consolidated at $7.71 per share. At the time Mr. Teague commenced his onslaught Goldfield Consolidated shares had slumped from $10 to $7.50. Mr. Teague alleged that the market on the stock was being juggled and speculators were being milked. Mr. Baruch, he asserted, had sold the stock down to $7.50 per share on the strength of his option, and was now tempted to break the market, sell the stock short and cover all at much lower prices.

Within two weeks after the publication of Mr. Teague's exposé of the terms of the outstanding option to Mr. Baruch, Goldfield Consolidated shares dropped to under $6. The story evidently had its effect.

The issue of the paper which chronicled the break to $6 contained an editorial headed "Nixon in the Rôle of Brutus." It demanded of Senator Nixon that he stand behind the stock and support the market, and also called upon him to declare the payment of dividends which he had promised to stockholders in his annual report dated two months prior.

People in Nevada began asking, "Who is Teague?" Mr. Teague caused the publisher of the *Nevada Mining News*, who was Hugh Montgomery, formerly business manager of the Chicago *Tribune*, to explain over his signature that Mr. Teague had been the political editor of the Baltimore *American*, later an editorial writer for the Philadelphia *North American*, and that while on the Philadelphia *North American* he had crusaded against get-rich-quick swindlers who had headquarters in Philadelphia, with the result that the Storey Cotton Co., the Provident Investment Bureau, the Haight & Freese Co. and other bucketshop concerns were put out of business. On evidence furnished by him, it was stated, Mr. Teague secured the conviction by the United States Government of Stanley Frances and Frank C. Marrin as chief conspirators in the $400,000 Storey cotton swindle. Finally, the article said, Mr. Teague was engaged by a far-famed magazine to expose bucketshop iniquities in the United States. This series of articles had appeared in 1906.

The biographical sketch seemed to satisfy readers that they were getting their dope straight on Goldfield Consolidated. My name at this time did not appear in connection with the publication except as part of the aggregation of Sullivan & Rice who advertised therein, but I was openly accused by Messrs. Nixon and Wingfield of dictating the policy of the paper. This was a half-truth. My sympathies were with the stockholders of Goldfield Consolidated—that's all.

The story is told in Nevada that when Senator Nixon received the check for $1,000,000 from Berney Baruch, after having executed notes of the Goldfield Consolidated, signed by himself as president and endorsed by him as an

BEATTY, NEV.   190   NO.

BULLFROG BANK AND TRUST COMPANY

PAY TO   OR ORDER $

                                     DOLLARS

...The... First National Bank of Rhyolite
CAPITAL, $50,000.00

Officers

OSCAR J. SMITH, PRESIDENT
BERT L. SMITH, VICE-PREST.   F. H. STICKNEY, CASHIER
T. A. FLEMING, VICE-PREST.   M. E. STICKNEY, ASS'T CASHIER

Directors

C. E. BRYSON
P. A. BUSCH
T. A. FLEMING
BERT L. SMITH
S. F. LINDSAY
MALCOLM L. MACDONALD
JAMES McENTEE
L. P. McGARRY
L. O. RAY
F. H. STICKNEY
OSCAR J. SMITH

NO. 8686

Auto in front of Southern Ho[tel] in Rhyolite

individual, he took luncheon at the Waldorf-Astoria in New York. When the waiter presented the bill the Senator ostentatiously tendered the $1,000,000 check in payment. The waiter put it all over the Senator by politely stating that if he wished to pay his dinner check out of the proceeds, Proprietor Boldt would undoubtedly attend to the matter for him. The Senator was forced to tell the waiter he was only joking.

The *Nevada Mining News* appeared to be catching on and was now printing 28,000 copies weekly. Sample copies were sent in every direction with the idea of acquainting investors with its existence.

A day after the issue appeared containing the editorial in which Senator Nixon was accused of playing the rôle of Brutus, I was stopped on the street by the editor of the Reno *Gazette*, a newspaper which is loyally attached to the Senator and his friends.

"The Senator wants to see you, Rice. Better go over to the bank right away. If you know what's good for you, you'll do it," the *Gazette* man said.

"I will, like ———!" I replied. "My office is up in the Clay Peters Building, and if the Senator has anything to say to me he can give me a call. I am not one of his sycophants, and I am not going." I didn't go.

An hour afterward the editor of the *Gazette* met me again. "Senator Nixon wants to see you at his office right away," he said bluntly.

"About what?" I inquired.

"About articles which have appeared in the *Nevada Mining News*," he answered.

"Very well," I replied, "I'll send the editor over."

Turning to Mr. Teague, I said, "I have no business with Senator Nixon, and if he has anything to communicate regarding the newspaper you, the editor, are the man for him to say it to."

Mr. Teague went over to the Nixon National Bank and entered the directors' room. My stenographer accompanied him as far as the door and took a seat outside, in the banking room.

As Mr. Teague entered, Senator Nixon jumped to his feet. He looked black as thunder. He quivered with rage.

"Why don't Rice come over here himself, eh? He daren't! I've got his record from boyhood jacketed in these drawers. While I have not read it, I know the story, and I am going to have it published in a bunch of newspapers so the world can know who is holding me up to public scorn!" the Senator spluttered.

In relating what transpired Mr. Teague later informed me that the Senator's wrathful indignation appealed to him as so grotesquely comic he felt like laughing, but he thought it a poor newspaper stunt to incense him further at a moment when it looked as if, by appeasing him, he could tempt him into volubility. Soon Mr. Teague had the Senator at ease, pouring forth a long interview, full of acrimony and affectation, which Mr. Teague promised to publish in the *Nevada Mining News*.

Mr. Teague reported to me that the Senator construed his pacifying attitude as meaning that I would undoubtedly listen to reason and that his threat would most certainly accomplish its purpose.

CALLING FOR A SHOW-DOWN

WHEN MR. TEAGUE finished narrating to me what had transpired I was beside myself. Presently I gave him these instructions: "Write out the interview with the Senator. Have two carbon copies made. When finished, take the three copies over to the Senator and have him read them and put his O.K. on them. After you have done that, give the Senator one copy, give the printer a copy, and put the other copy in the safe. As soon as the copy of the interview is in the printer's hands, sit down and write an editorial. Head it 'A United States Senator with a Blackmailing Mind.' Publish my record in full. Tell of everything of any consequence I ever did, good or bad. Parallel my record with the Senator's record. Tell the people of Nevada all the facts about the Senator's threat. Say to them nobody can blackmail me, and ask them to choose between us."

On May 25, 1907, the editorial, headed "Nixon a Senator with a Blackmailing Mind," appeared. It was a passionate denouncement, calculated to stir the blood. Also there appeared Senator Nixon's interview in full.

In the interview the Senator had made an effort to disentangle himself from a seemingly inextricable network in which he was enmeshed, and the paper contained still another editorial lambasting him in amplitude for trying to practise on the credulity of the newspaper's readers. The editor accused him of equivocation, artful dodging, false coloring, exaggeration, suppression of truth, cupidity and knavery.

The arraignment wrought an undoubted sensation. The effect on the Nevada public was unmistakable. It reminded me more of the motionless and breathless attitude of an audience at the third act climax of a four act drama, than anything else. The Senator was not seen on the streets of Reno for two months afterwards. For a fortnight afterward he didn't even call at the offices of the bank. When he did finally resume his visits to the bank he came in his automobile. He was whisked to the door of the building, immediately secreted himself in the directors' room and was not get-at-able.

The Nevada Mining News

Has been right on every market turn the last 15 months.

Its 40,000 readers have recently piled up tremendous profits in Nevada issues.

The big July-August bulge in Nevada was forecasted in the Nevada Mining News two weeks before it happened, just as the terrific slump that had preceded it was foretold.

It is the recognized authority in Nevada, San Francisco and New York on Nevada mining securities. This is conceded by everybody.

The Nevada Mining News
Dares to Tell the Truth

It employs a trained force of high salaried correspondents in every Nevada mining camp and in every market where Nevada shares are traded in.

It pays more money for quick news of the mines and the markets than any other newspaper in the United States, and it pays bigger telegraph tolls in gathering news than all other mining publications.

Its news and market prophecies are so accurate, it is now conceded to have as great an individual following as any other financial newspaper in the United States or Europe.

It is published every Thursday. Subscription rates, $5 per year.

The Nevada Mining News

Geo. Graham Rice
Editor
Reno .. Nevada

Editor Rice claimed 40,000 readers

Leading citizens, including the directors of a number of banks in Reno, made clandestine calls at my office, shook my hand, felicitated me over the stand I took, and went away. Even George Wingfield, the Senator's partner, it was reported (and I afterward corroborated this from the lips of George Wingfield himself), backed me up in the stand I had taken. The general sentiment in the state appeared to be that the threat was a lowdown trick, and that of the two I had the less to be ashamed of.

When the Senator read the article "Nixon a Senator with a Blackmailing Mind" it is said he telegraphed to former Governor Thomas of Colorado, his counsel, and asked him to come to Reno.

"If I don't say something in answer to this awful attack, I'll choke!" cried the Senator as he nervously walked the floor.

"Did you sign that interview which they published?" asked Governor Thomas.

"Yes," said the Senator.

"Well, then, if you say anything at all now, they'll choke you," answered Governor Thomas.

During the course of our attacks on Senator Nixon in the *Nevada Mining News* which followed at various intervals, the newspaper accused him of making promises of early dividends to Goldfield Consolidated stockholders which he knew he could not keep; of having been the state agent in Nevada of the Southern Pacific Co. at $150 per month during the Huntington régime when legislatures were bought; of having bilked the investing public out of millions in Goldfield; of having carved his fortune, that made possible the acquisition by him and his partner of control of the Goldfield Consolidated, out of a gambling house in Tonopah; of having gathered his first mining property and mining stock interests in Goldfield from prospectors who lost money and surrendered their mining claims and stock certificates to the gambling house in lieu of the cash; and of being generally a financial and political freebooter of the most despicable sort. And the Senator never sued for libel nor proceeded in the courts in any way whatsoever to obtain a retraction.

MANIPULATING GOLDFIELD CON

About a week after the publication of the editorial headed "Nixon a Senator with a Blackmailing Mind," when Goldfield Consolidated stock had slumped to around $7, the *Nevada Mining News* in big bold-faced type urged its readers to place their buying orders for Goldfield Consolidated at $4 a share, saying that New York mining stock brokers advised their clients that the stock would almost certainly go down to that figure because of the Senator's mistakes in the financial management of the company. That edition contained another editorial on Senator Nixon, headed "Branding a Bilker." It accused him of saying in his annual report a few months previous that payments of dividends on a regular basis would commence within a short time, and contrasted this statement with the signed interview published in the *Nevada Mining News*, in which he said dividends would be paid "whenever the trustees thought it wise to do so and not before."

Within a day thereafter the stock busted wide open to $5⅛ bid, $5¼ asked, and the whole Goldfield list smashed farther in sympathy. By June 8th Goldfield Consolidated had crashed to $4.50.

On the dip from $7.50 to $4.50 an opportunity had been offered to Berney Baruch and his associates to buy back in the open market all of the stock they might have sold on the way down from $10 to $7.75, which was the option price. Then the stock was promptly manipulated back to $7. On the way back to $7, the outstanding short interest (of other traders who had accompanied the decline with their selling orders) was forced to cover.

To help along the covering by outsiders up to the $7 point a report was circulated by lieutenants of Senator Nixon in Reno that a dividend would be declared before the end of June, and almost simultaneously the general manager of the mining company in Goldfield put forth a similar tip. As the market began to recover toward the $7 point, Senator Nixon went to San Francisco and was seen often at the sessions on the floor of the San Francisco Stock and Exchange Board. On the day before the bulge to $7 he was quoted in a San Francisco newspaper as saying that Goldfield Consolidated was such a good thing he would not take $20 per share for his stock.

When the stock hit $7 and the shorts were being squeezed the hardest, Senator Nixon was quoted as saying in still another interview that a dividend was not far away. This interview was carried over the telegraph wires to all market centers by the Associated Press. At the same time a story was printed in the New York *Times* saying that it was reported on the Street that J. Pierpont Morgan, acting for the Baruch-Ryan crowd, had taken over the control of the Goldfield Consolidated. The shorts were successfully driven to cover. Then the price eased off again in a day from $7 to $6⅛.

A month later Mr. Teague became editor in chief of the *Nevada State Journal* and severed his connection with the *Nevada Mining News*. I succeeded Mr. Teague as editor and my name appeared at the head of the editorial columns. At about the same time the Sullivan & Rice enterprise was abandoned. I discovered that most of the money Mr. Sullivan had put into the corporation had been borrowed by him from a member of my own family with whom he had hypothecated most of his stock in the company. A rumpus ensued which ended in the shutting up of the shop.

By August Goldfield Consolidated had been manipulated back to $8.37½ a share. Mr. Baruch's option could certainly prove of little value to him unless the stock sold higher at periods than $7.50. But he now evidently found it a hard job to hold the stock above $7.50. By September it had receded again to $7.40. At this period it was reported in Reno that George Wingfield, sick of his partner's bad bargain, was beginning to assert himself and demanded that the Baruch option be cancelled at whatever cost.

The erratic price movement of the stock was causing the loss of public confidence. The manipulation appeared to be raw. Without any important transpiration except the news of the Baruch option and the varying statements put out by Senator Nixon from time to time regarding the plans of the company, which was now awaiting the erection of a huge mill before going on a regular producing basis, the stock had dropped from $10 to $4.50, recovered to $7 and eased off to $6⅛, rallied to above $8, and was again tumbling.

The option to Mr. Baruch was conceded to be practically a flat failure from a company standpoint, only 20,000 shares of stock having been purchased by Mr. Baruch from the treasury of the company in nine months. The impression prevailed that Mr. Baruch was milking the market and held the option principally as a club to accomplish his market designs. Moreover, nearly every broker, investor and speculator residing in Goldfield by this time had gone broke because of the vagaries of this stock in the market, and the losses in bad loans and unsecured overdrafts incurred by John S. Cook & Co.'s bank, controlled by Messrs. Nixon and Wingfield, was said to total nearly $2,000,000 as a result of the almost general smash in market values. The entire Goldfield list, with the exception of Goldfield Consolidated, was now selling at 25 cents on the dollar compared with boom prices of less than a year before, and it was a rather ordinary piker sort of broker or speculator in Goldfield who at this time could not boast of being in soak to John S. Cook & Co.'s bank anywhere from $15,000 to $100,000.

On September 23 the Goldfield Consolidated directorate met at Goldfield. After the meeting it was officially announced that the option held by Mr. Baruch on 1,000,000 shares at $7.75 had been canceled and that Mr. Baruch had been given sufficient of the optional stock to liquidate the $1,000,000 obligation of the company, leaving the company free of debt and with a cash reserve of nearly $2,000,000. It was stated that Mr. Baruch had originally been given the option for services in securing the loan of $1,000,000 from J. Kennedy Todd & Co. of New York for 13 months with interest at the rate of six per cent, and that the price of $7.75 was an average one, indicating that Mr. Baruch held an option on stock at varying figures on a scale up from a considerably lower than $7.75, which he might have exercised in whole or in part.

It was also disclosed that a large block of Goldfield Consolidated stock had been put up as collateral for the note. Because the officials of the company declared by resolution that the unused certificates shall be canceled, it was generally believed that the entire 1,000,000 shares under option to Mr. Baruch, had been put up as security.

The official statement of the company said that the option had been turned back to the company on a satisfactory basis. No figures were given out. Dispatches from San Francisco to the *Nevada Mining News*, which I promptly published, alleged that Mr. Baruch was given 200,000 or more shares of Goldfield Consolidated in settlement of the loan to the corporation of $1,000,000 and for the surrender of his option on 1,000,000 shares at an average price of $7.75.

The 200,000 shares of stock was taken out of the collateral at the rate of $5 per share on a day when Goldfield Consolidated was selling around $7.50, after the stock had been manipulated to a fare-ye-well and against a market price of $10 for the stock on the day the option was given.

No denial was ever published. My opinion, based on private investigation and on analysis of the company's reports, is that Mr. Baruch fared even better than as outlined above.

The giving of the option had made it dangerous for anybody except Mr. Baruch to attempt to hold the stock above $7.75 per share after the option had been given, and the company in addition was now mulcted for the difference between the low price per share at which settlement was made with Mr. Baruch and the price at which the stock could have been sold had it been quietly disposed of on the market during the period of nine months which had preceded the date of cancellation. As a matter of fact, there was no necessity at all for settling the loan with stock, the company having in its treasury more than sufficient to repay the loan, and the money was not due. The real purpose, apparently, was to shroud in darkness the exact amount given to Mr. Baruch to release the company from the option and to keep Messrs. Nixon and Wingfield's Goldfield bank, which was the depositary of the mining company, in funds.

Instead of quieting the stockholders the surrender of the option again thrust into the limelight the entire transaction and proved to be an exacerbation.

The immediate effect was that Goldfield Consolidated began to slump again, and in a few days sold down to $6.50. From this point it kept on tobogganing during a period of weeks down to the $3.50 point—a depreciation in market price for the capitalization of the company, within a year of its promotion at $10 a share, of $23,400,000—before rallying once.

ENTER, NAT C. GOODWIN & CO.

A MINING PARTNERSHIP between Nat C. Goodwin, the actor, and Dan Edwards had been formed at Reno a little before this time. Dan Edwards was a hustling young mining man who had engaged in the business of turning properties to promoters. In August, when Goldfield Consolidated was selling around $7.50, Mr. Edwards had asked me to give him a good market tip. I told him to sell Goldfield Consolidated short.

When it hit $6.50 around October 1st he saluted me thus, "Got to hand it to you. I have been trying to make my new firm stick, but it don't seem to work. I guess I don't know how to handle the situation in times like this. How would you like to join us?"

"How much capital have you got?" I asked.

"Five thousand of Nat's money," he answered.

"Get another man with $5,000," I said, "and I'll talk to you."

A young Easterner engaged in mining, named Warren A. Miller, was stopping at the Riverside Hotel. Within an hour Mr. Edwards had him lined up. A week later Nat C. Goodwin & Co. was incorporated with Nat C. Goodwin president, Mr. Miller vice-president and general manager, and Dan Edwards secretary. The new corporation engaged to give me a salary for showing it how and an interest for other substantial considerations.

Within a fortnight the corporation of Nat C. Goodwin & Co. was making money, not as promoters, however, but as demoters. Instead of at first promoting a mining company and earning its profits on the constructive side of the market, it turned the tables and made money on the destructive side—of Goldfield Consolidated.

During the first half of 1907 I had felt the country's speculative pulse from day to day with the promotion literature of the Sullivan & Rice corporation. Although its new mining company, the Rich Gulch Wonder, had boasted of a very high class directorate and the property was conceded to have merit, the public refused to enthuse. Instead of subscribing for large blocks, scattering purchases had been made, and money in dribs and drabs had been grudgingly paid over. The Wonder mining camp boom had died

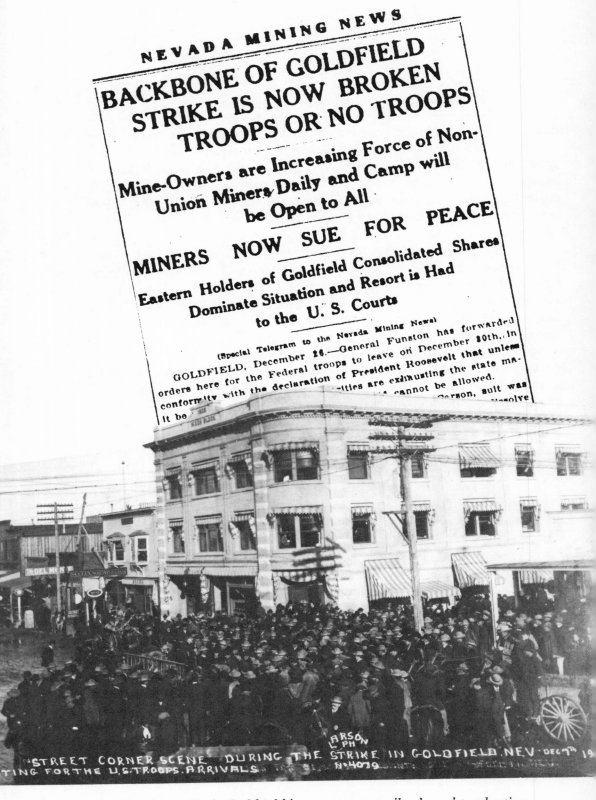

The big labor strike in Goldfield in 1907 temporarily slowed production

abornin'. Investors seemed to have had enough of mining stock speculation for a while.

Prices of listed Nevada issues were crumpling like seersucker in the rain. By this time the awful mess that had been made of Goldfield affairs through the mistakes of Messrs. Nixon and Wingfield had resulted in a depreciation in market value of more than $100,000,000 in listed Nevada issues. This in itself was sufficient to kill a world of buying sentiment.

You have to be a rainbow chaser by nature to be a successful promoter, but even I, despite my chronic optimism, began to feel the influence of what was transpiring. I made a flip-flop and turned bear on the whole market.

On October 17th the Heinze failure occurred in New York. Five days later the embarrassment of the Knickerbocker Trust Co. was announced. I glued my ears to the ground.

Nat C. Goodwin & Co. shorted the mining stock market so far as its limited capital would permit. On the day Mr. Heinze went overboard the company was already short 2,000 shares of Goldfield Consolidated at around $6. On hearing that the Knickerbocker Trust Co. was in trouble it promptly shorted 2,000 shares more at a lower figure.

On the afternoon when the news reached Reno of the Knickerbocker Trust Co.'s embarrassment I received a private telegram from Chicago stating that the paper of the State Bank & Trust Co. of Goldfield, Tonopah and Carson City had gone to protest in San Francisco. This set my blood tingling. I knew that meant a general Nevada bust.

Next morning Nat C. Goodwin & Co. shorted 2,000 shares more of Goldfield Consolidated at about $5⅛. Later in the day the failure of the State Bank & Trust Co. was announced. A run followed on the Nye & Ormsby County Bank and its branches in Reno, Carson City, Tonopah, Goldfield, and Manhattan, and in two hours that institution, too, closed its doors.

Goldfield Consolidated promptly broke to $4 a share. Around this point Nat C. Goodwin & Co. covered its short sales, at discretion.

All of the Nixon banks in Nevada experienced runs as a result of the failure of the two Nevada banking institutions. So did the other banks.

Governor Sparks was appealed to by Nevada bank officials between two suns to come to the rescue. Without hesitation he declared a series of legal holidays to enable the banks of the state which were still standing on their feet to catch their breath. These banks finally threw open their doors, but when they did, those of Reno met depositors' withdrawals with asset money instead of legal tender. The only bank in Reno which had refused to take advantage of the enforced legal holidays was the Scheeline Banking & Trust

Co. And when asset money was finally resorted to as a makeshift, M. Scheeline, the president, was made custodian of the bonds which were put up by the associated Reno banks to secure payment. This restored confidence.

It was believed in Nevada at the time of the failure of the mining camp banks, the State Bank & Trust Co. and the Nye & Ormsby County, that the Nixon institution in Goldfield would have found it hard to weather the storm but for the fact that the Goldfield bank was believed to have upward of $2,000,000 of the Goldfield Consolidated Mines Co.'s money on deposit.

When the State Bank & Trust Co. went to the wall Senator Nixon, in an interview published in his Reno newspaper, charged the failure of the State Bank & Trust Co. to me. He alleged that the State Bank & Trust Co. lost $375,000 by the failure of the Sullivan Trust Co. ten months before, and that I had broken the bank. The liabilities of the bank were $3,000,000, and its Sullivan Trust Co. loss was only a drop in the bucket. The Senator didn't fool anybody, not even himself. His effort was an ill-concealed attempt to prepossess the public against me, and was received by Nevada people as such.

Senator Nixon indulged in some more interview with a view to stemming the tide of liquidation in Goldfield Consolidated. Notwithstanding the fact that the company had only recently resorted to the sale of treasury stock for money raising purposes, he asserted that a quarterly dividend, payable January 25th, would probably be declared. Beyond a question this statement was made for market purposes at a time when the Senator was sweating money-blood.

The stock promptly tobogganed farther on the strength of the dividend forecast. The Senator's interviews had now become a standing joke in the community. Speculators and brokers had learned the wisdom of "coppering" anything the Senator said.

THE STORY OF THE GOLDFIELD LABOR "RIOTS"

A LARGE FORCE of miners was discharged from the Goldfield Consolidated properties. The action of the company in laying off its men at such a distressful period was denounced. It was alleged that Senator Nixon's Goldfield bank could not afford to pay out the money on deposit to the credit of the company because it was required for bank purposes. The money was apparently being hoarded during the money stringency to help the bank out of a tight place. After events appeared fully to confirm this theory.

Right in the teeth of the panic, during the depressed and troublous days of the latter part of November;—when current finance was deeply affected; when Goldfield Consolidated was selling below the $4 point and the entire

ROOSEVELT WILL WITHDRAW TROOPS FROM GOLDFIELD IN SHORT TIME SAYS A MESSAGE

Does Not Believe That Conditions And Facts Justify Retaining Troops and Original Calling of Soldiers

WASHINGTON, Jan. 22.— President Roosevelt has determined on withdrawal of the Federal troops from Goldfield, Nevada, shortly after the legislature begins its special session Tuesday.

This intention was made known at the White House today when the report of the special investigating commission was made public, together with a letter from the Governor Sparks. The Presi...

The operators of the Goldfield district have been instrumental in obtaining federal aid and in bringing troops to Goldfield. On the attorneys for the commission they in re... their case...

Governor John Sparks, friend of George Graham Rice, called for troops to quell the Goldfield riot

STORY OF SITUATION FROM UNION STAND

The Other Side Is Presented by a Man Who Is Interested and Has Interests in Southern Camp

(Special Correspondence.)

Goldfield, Jan. 13.—The mountains of Nevada may be full of gold; the very bowels of the earth may be pregnant with it, but until it is taken out it is of no use to man. Two elements are necessary for its extraction, capital and labor, and they must co-operate. One would think that in an industry in which it can be compensated, namely, the coin of the realm, no difficulty could possibly arise between capital and labor as to the manner and method of making that compensation. Here is no variance in the market price of the product, no change in the demand. The only exigencies that can affect the wage scale are those that have to do with the cost of pr... The greatest hazard, name... product... eliminate... consider... is the co... concede... an insur... by agre... wage s...

The... years... the wa... seen i... gold... that... bition... reaso... man... both... that... oth... in... los... is... n... h... c... a...

are addressed...

On November 1, 1907, ... peace between employers and employees in Goldfield. On or about that time Manager Seibert of the Mo... lease telephoned

paid in cash out of the product of the mines." The union approved the recommendations of the president. Mr. Wingfield was asked to fix a time for a meeting of the operators and bankers with the union committee. He suggested that the union submit the matter in writing to the secretary of the association, Mr. Erb. This was done, and Mr. Erb stated that the operators and bankers could not meet the committee of the union, but had prepared a statement, which is as follows: "A special meeting of the mine operators was held for the purpose of considering the recommendations presented in your communication. After due deliberation on the part of the members of the association, the following resolution was passed: 'Resolved, that all of the companies op... in the Goldfield mining dis... bank checks

bankers will be as f... panies employing men in this district will pay for labor in checks payable in exchange at the local banks.

operators immediately took the position stated in their correspondence, and announced that they would not recede from it. The following questions were asked Mr. Murphy, the only real mine owner present:

Q. Will you back up any paper you may issue during this panic with the product of your mine? A. No, I will not. Q. Will you back up any paper you may issue with the mine itself? A. No, I will not. Q. Will you agree to leave the words "payable in exchange" from the checks if the miners will agree not to demand real money until you are in a position to pay it? A. No, I will not. This was practically all that came of the meeting, although upon its breaking up the operators announced that they would further consider the matter. On the same day the operators, after acknowledging the receipt of the la... letter from the union, addressed ... the following: "The Goldf... association wish... a reconsideration... issue with your c... iation offers the... xplanation of its... November 18th. ors cannot off... dition to that o... banks for ca... y the local bank... t place the ore s... ct as a guaran... ks issued by th... reason that mo... the ore shipped... cal banks and the... rect to the loca... ks do not issue... t against collater... n the camp, and... ssibly issue ch... less secured by... e local banks w... to any operate... which is nego... Sunday mornin... 07, the miners q... issuing only s... work.

... is to be noted ion did not at any time i... The order. The men simpl... banks have agreed to and will work where scrip was pa...

Nevada share list had suffered an average depreciation of about 85 per cent from the highs reached during the Goldfield boom of the year before; when the state of Nevada was racked from end to end by the serious losses incurred by citizens through the failure of the Nye & Ormsby County's and the State Bank & Trust Co.'s chain of banks, totaling nearly $6,000,000, and it appeared that the credit of the State had already been shattered almost beyond repair—a fresh blow was administered.

Government troops were reported to be enroute to Goldfield from San Francisco to preserve the law. It had been represented to the President of the United States that Goldfield was in a state of anarchy. Goldfield wasn't. As a matter of fact, the situation in Goldfield with the miners, from the standpoint of law and order, was never good, but it was as good then as it had been in eighteen months. True, there had been some lawlessness, but no riot, and the sheriff of the county had made no call whatsoever on the Governor for any aid.

During the first days of the panic Nixon and Wingfield's Goldfield bank, John S. Cook & Co., had tendered the miners the bank's unsecured scrip in lieu of money for the payment of wages. The miners refused acceptance. They were willing to take time checks of two, three or four months, bearing the mining company's signature, but balked at the idea of becoming creditors of the bank.

It has been stated to me by a number of Goldfield brokers who were present in the camp at the time that the miners had even decided to concede this point, when an outsider secured by intrigue and money sufficient voting power at a meeting of the executive committee of the Miners' Union to pass a resolution objecting to the bank's scrip. The refusal to accept the bank's scrip was at once made an excuse by the Goldfield Mine Owners' Association, which was dominated by George Wingfield, to determine upon a lockout and simultaneously to demand Federal intervention.

If Messrs. Nixon and Wingfield's bank needed money, as the tender of unsecured scrip indicated all too plainly, the complete shutdown which left with the bank as available resources approximately $2,000,000 in the account of the Goldfield Consolidated Mines Co., was a perfect stop-gap; and the need of the presence of troops was a fine coincidental excuse for the shutdown. Incidentally, it would rid Goldfield of the Miners' Union, which voted to a man against Senator Nixon's Republican candidate for office, and would permit the importation of foreign labor, an expedient which was afterward successfully resorted to.

Senator Nixon brought pressure to bear at Washington. He invoked the good offices of Uncle Sam and urged that federal troops be sent to the state. He was assisted by Congressman Bartlett in laying the matter before the departments. The wires between Goldfield, Reno, Carson City and Washington were kept hot with an interchange of views. President Roosevelt finally informed the Senator he could not send the soldiers unless the Governor of Nevada wired that a state of anarchy actually existed which the state itself was powerless to put down.

Governor Sparks, honest as the day was long and unsuspecting of any trickery or jobbery, listened to a Goldfield committee and permitted a dispatch to be sent to Washington over his signature representing that such conditions existed.

Thereupon Brigadier-General Funston, at the head of two thousand troops was ordered to Goldfield. The state being without any militia and the representations made by Governor Sparks in his dispatches being strong on the point that a state of anarchy actually existed in Goldfield, the President finally succumbed.

The maneuver was as swift as it was unexpected. Nevada people at first could not understand what it was all about. Dispatches from Goldfield to Reno said the town was quiet. The nearest approach to an overt act of recent occurrence that had been chronicled was the alleged theft a few days before of a box or two of dynamite, about 300 feet of fuse and a quantity of caps that were said to have been clandestinely removed from the Booth mine in Goldfield. The theft, if theft there was, was charged to the miners, but proof was lacking.

On the arrival of the troops in Goldfield the Goldfield Consolidated announced a new wage scale, reducing the miners' wages from $5 to $4 and in some cases from $5 to $3.50. This was a new move, calculated to rouse the ire of the wage workers and to prolong the lockout. Messrs. Nixon and Wingfield's bank in Goldfield announced at the same time that it would thereafter discharge all of the payrolls of the company in gold. But there were no payrolls of any consequence then, as the mines had been shut down and the work force locked out.

General Funston on his arrival in Goldfield interviewed mine operators, union miners and citizens generally with a view to determining the necessity for maintaining government troops there. He discovered that the administration had been buncoed. The General wired the President his opinion. President Roosevelt quickly dispatched a commission to Goldfield to conduct a public inquiry. This commission consisted of Charles B. Neal, Labor Commissioner; Herbert Knox Smith, Commissioner of Corporations, and Lawrence

WHEN GOLDFIELD CONSOLIDATED HIT $5 PER SHARE TODAY IT REACHED
BOTTOM. IT WILL NOW REACT. BUY IT QUICKLY. ALSO BUY EAGLE'S NEST
FAIRVIEW. BUY BOTH OUTRIGHT. DON'T BUY THEM ON MARGIN. AVOID
THE SHAKING OUT PROCESS. — George Graham Rice

With the great Malpais Mesa serving as backdrop, the one year old Combination Mill with its 10 stamps crushed about 80 tons of ore daily. In March 1907 the Goldfield Consolidated acquired this property.

O. Murray, Assistant Secretary of the Department of Commerce and Labor. They heard testimony day and night for a week.

They reported to President Roosevelt that there was no occasion for the presence of troops in Goldfield and that the statements telegraphed to President Roosevelt by Governor Sparks, indicating the existence of a state of anarchy, were without justification. The report was given to the Associated Press and received wide publicity. The President also issued a broadside backing up the findings, and its message was telegraphed far and wide.

Eastern editorial writers poured out torrents of abuse on Governor Sparks. Senator Nixon went unscathed.

THE DEATH OF GOVERNOR SPARKS

FEELING UNDER A WEIGHT of obligation to Governor Sparks, who had headed nearly all of the Sullivan Trust Co. promotions as president, I tried editorially in the *Nevada Mining News* to justify the Governor's action. But it was a wee voice drowned in an ocean of adverse opinion and was entirely without echo. It didn't even soothe the Governor.

The Governor, honest, simple old man, broken in purse, in health and in spirit, grieved over the President's denouncement, took to bed, and died of a broken heart.

At his imposing funeral pageant in Reno, which was attended by thousands of mourners, who had come from all parts of the state to pay homage to the grand old man and who followed the hearse to the cemetery, Senator Nixon and his partner, George Wingfield, were conspicuous by their absence.

Even at the moment when the grave closed over his remains the troops were leaving Goldfield.

"It's the Death March," said one of the bereaved.

The bringing of federal troops to Goldfield accomplished its purpose. The Miners' Union was destroyed and sufficient time was gained to enable the financial atmosphere to clarify. By the time the troops departed, Goldfield Consolidated had rallied to $5 per share. The panic was over. Money was comparatively easy again.

I ask the reader's indulgence for having devoted so much space to the facts bearing on the appearance in Nevada of United States troops at a time when there was no valid occasion for their presence. I feel that it is an important chapter of my experiences and is fraught with interest to the general reader, because it illustrates how easy it is to direct the powerful machinery of our great government so as to carry out the machinations of evil-minded men.

You might think, after this demonstration of the lengths to which Senator

Nixon went to accomplish a set purpose, and after witnessing the success which attended his efforts, that a poverty stricken individual like myself, who had had the hardihood to conduct a newspaper campaign in the Senator's own home town against his financial and political activities, would judge it the better part of valor to emigrate from the state. Well, I didn't. I stayed right on the spot. That I would hear from Washington later I had no doubt, but I stuck, just the same, until my business interests called me away.

I wasn't wrong in my deductions. Within a few months thereafter there came a visit to my office in Reno by a postal inspector, who was apparently sicked on to the job, and but for the quick intercession by telegraph of United States Senator Francis B. Newlands of Nevada with the Postmaster General at Washington, I am certain that potent influences even at that early day would successfully have started something. But of this more at another time. I am ahead of my story.

W. H. Knickerbocker, who delivered the famous eulogy to Riley Grannan (see pages 236-237), is shown surrounded by his prospecting pals in Goldfield, in 1903

7. Birth of Rawhide

BECAUSE RAWHIDE, the new Nevada gold camp, was born during the financial crisis of 1907, I couldn't see any future ahead of it from the promoter's coign of vantage—not through a pair of field glasses. It requires capital to develop likely looking gold prospects into dividend paying mines, and I could not imagine where the money was going to come from. Eastern securities markets were in the doldrums. Time money commanded a big premium. Prices for all descriptions of mining stocks had flattened out to almost nothing. Investors were at their wits' end to protect commitments already made. Financiers everywhere were depressed. A revulsion of sentiment toward speculation had set in, seemingly for keeps. Only a hair-brained enthusiast of the wild-eyed order could hope at such a time possibly to succeed in the marketing of new mining issues.

A financial panic has no terrors, however, for gold seekers. The lure of gold is irresistible. Money stringency serves only to strengthen the natural incentive. By the first week in January 1908, fully 2,000 people were reported to be in Rawhide. At the end of January the population had grown to 3,000. The camp easily held the center of the mining stage in Nevada.

Many of the Rawhide pioneers hailed from Tonopah and Goldfield. Without exception the opinion of these veterans appeared to be that the surface showings of the new district excelled those of either of the older camps. Never before in the history of mining in the West had there been discovered a quartz deposit so seemingly rich in the yellow metal at or near the surface which at the same time embraced so large an area of auriferous mineralization. Goldfield, at the same early age, had been a mere collection of prospectors' tents, while Rawhide was a thriving, bustling, populous camp with more than a hundred leasing outfits conducting systematic mining operations.

Booming Rawhide looked like a movie set town in 1908

Page Four

Nevada Mining News
The Financial Mining Newspaper of the West

**PUBLISHED EVERY THURSDAY BY THE
NEVADA MINING NEWS BUREAU, Inc.
RENO, NEVADA**

GEORGE GRAHAM RICE . . . . . EDITOR
SAM C. DUNHAM, Mining Editor

SUBSCRIPTION RATES:

Per Annum $5.00
Six Months 2.50
Three Months 1.25
Single Copies .... .10

ADVERTISING RATES: 35 cents per agate line ($3.50 per inch). No discounts for time, cash or space. No commissions allowed to advertising agents. Terms: Cash with order.

Address all communications to the "Nevada Mining News Bureau," Reno, Nevada.

Circulation, 30,500

Vol. 1. RENO, NEVADA, THURSDAY, FEB. 13, 1908. No

Sam C. Dunham, for the past six years editor and part owner of the Tonopah Miner, the best-posted, most conservative and able writer on mining topics in Nevada, has joined the forces of NEVADA MINING NEWS and will hereafter preside over the mining news department.

THE RUSH TO RAWHIDE

The "rush" to Rawhide, which is now in full swing, is the greatest stampede in the history of gold mining. People are pouring into the new camp at the rate of 100 a day and the population of the district is, at the present writing, with every indication that within exceed that of Goldfield. When it is considered that the movement had its inception at a time when Nevada was paralyzed by the money stringency while and that the district has grown and developed portions in the face of the worst and most trying depression this country has experienced for many years, unquestionably stands forth as the most in the world.

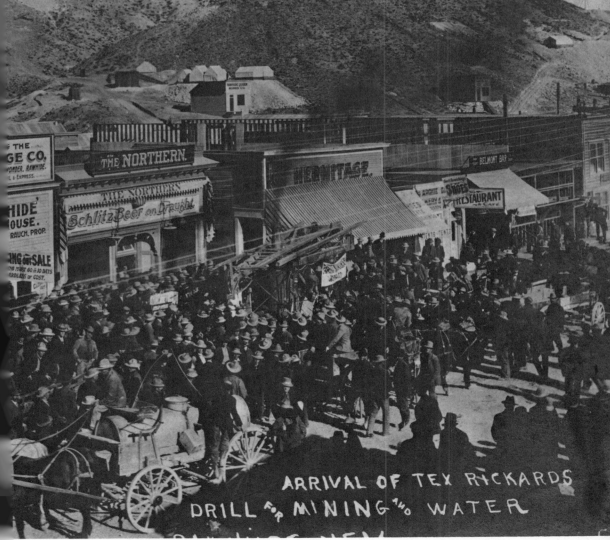

ARRIVAL OF TEX RICKARDS DRILL FOR MINING AND WATER

News was brought to Reno of a phenomenal strike made on Grutt Hill in Rawhide. Specimen rock taken from a seam of ore assayed $300,000 to the ton. The Kearns lease on Balloon Hill reported 15 feet of shipping ore on the 65-foot level which assayed from $300 to $500 to the ton. There was full verification of this. Also regular shipments were being made to the Goldfield reduction works.

Samples of rock were received in town that were studded with free gold. I was thrilled. Statements made by camp boosters that a part of Balloon Hill was "gold with a little rock in it," were not exaggerations, judging from the specimens that were placed in my possession.

My apathy began to melt away. Against my earlier judgment, I now began to change my attitude. The camp looked like the real thing, panic or no panic.

Why should not the American public, even in these tough financial times, enthuse about a gold camp with possibilities for money making such as are offered here, I asked myself. Don't drowning men grasp at straws? Is it not the habit of horse race players when they lose five races in succession to make a plunge bet on the sixth with a view to getting out even? This panic had impoverished hundreds of thousands. What more natural than that those who were hit hard should now fall over one another to get in on the good things of Rawhide? If the camp makes good, I reasoned, in the same measure that Goldfield did, early investors will roll up millions in profits.

I visited the camp. What I saw electrified me. Soon I was under the magic spell.

REAL GOLD AT RAWHIDE

HALF A DAY'S TRAMP over the hills seemed sufficient to convince anybody that the best of the practical miners of Nevada had put the stamp of their approval on the district. Most of the hundred or more operating leases of Rawhide were owned by these hard rock miners. More than half a dozen surface openings on Grutt Hill showed the presence of masses of gold-studded quartz. At the intersection of Rawhide's two principal thoroughfares a round of shots in a bold quartz outcrop revealed gold-silver ore that assayed $2,700 a ton. A gold beribboned dyke of quartz rhyolite struck boldly through Grutt Hill's towering peak. I walked along its strike and knocked off, with an ordinary prospector's pick, samples worth $2 to $5 a pound.

Across Stingaree Gulch to the south Balloon Hill's rugged hog-back formed a connection link between Grutt and Consolidated hills. The Kearns Nos. 1 and 2 leases on Balloon Hill were scenes of strikes of such extraordinary richness that they alone would have started a stampede in Alaska. The Murray lease on Consolidated Hill was rated as a veritable bonanza. There I saw quartz that was fully one-third gold.

Along the southern slope of Hooligan Hill several sets of leasers were mining ore so rich that guards were maintained through the night to prevent loss from theft. At the Alexander lease on Hooligan Hill the miners were crushing the richer quartz from their shafts and washing out gold to the value of $20 a pan.

These were the three principal centers of activity, but they by no means embraced the productive area of the district. Tall, skeleton-like gallows frames dotted the landscape for miles in every direction. The coughing of gasoline engines suggested the breathing of some spectral Titan in the throes of Herculean effort. I was forcefully impressed, too, with the class of miners at work.

It seemed to me there was no longer any room for cavil as to the fortune making possibilities of investors who put their money into the camp. Less than a half year old, Rawhide loomed up as the most active mining region I had ever seen at anything like the same age. It required nearly three years for Goldfield to make as good a showing, I reasoned.

During my earlier efforts at press agenting southern Nevada's mining camps I had to conjure in my mind's eye what the reality would be if half the hopes of camp enthusiasts were fulfilled. Here was apparently a fulfillment rather than a promise. At the threshold of the first stage of its development era Rawhide could boast of more actual producers and nearly as many operating properties as Goldfield could claim at the age of three years.

I recalled that Cripple Creek had been panic-born but had lived through the acute period of 1893-96 to take rank with the greatest gold camps of the world. I was more than convinced. Effervescent enthusiasm succeeded my earlier skepticism. History is about to repeat the record of Cripple Creek, I concluded.

THE RAWHIDE COALITION MINES COMPANY

GRUTT HILL, HOOLIGAN HILL, a part of Balloon Hill, and the intervening ground, forming a compact group of eight claims, 160 acres, were owned by a partnership of eight prospectors. The area formed the heart and backbone of the whole mining district.

I soon tied up this property for Nat C. Goodwin & Co. of Reno, with whom I was identified. A company, with 3,000,000 shares of the par value of $1 a share, was incorporated to take title. It was styled the Rawhide Coalition Mines Co.

Rice's actor pal,
Nat C. Goodwin

The August 1908 fire burst
Rawhide's promotional bubble

Rawhide publicity items

Of its entire capitalization, 750,000 shares were turned into the treasury of the Rawhide Coalition Mines. Nat C. Goodwin & Co. became agents for the sale of treasury stock, and were given an option by the company on 250,000 shares, to net the treasury $57,500 for purposes of administration and mine development. The Goodwin company also purchased 1,850,000 shares of the 2,250,000 shares of ownership stock, amounting to $443,500 more, or at the rate of 23.3 cents per share plus a commission of $12,500 to be paid to a go-between.

The ownership stock that was retained by the original owners, and the residue of treasury stock, amounting in all to 900,000 shares, were placed in pool.

When I made this deal the cash in bank of Nat C. Goodwin & Co. amounted to about $15,000. It was up to me to finance the undertaking. I did.

The contract I made called for only $10,000 in cash and the balance on time payments. Nat C. Goodwin & Co. didn't borrow money from any bank or individual, nor did anybody identified with the concern tax his personal resources to the extent of a single dollar to go through with the deal. The money was raised, first for the Coalition's treasury and later for the vendors, by appealing directly to the speculative instinct of the American investing public. The public, too, paid the expense that was incurred in reaching them. It did this by paying Nat C. Goodwin & Co. an advance in price on Coalition stock purchases, over and above the cost price.

Nat C. Goodwin & Co. had agreed to net a fraction more than 23 cents per share to both the treasury and the vendors without any deductions whatsoever. All of the advertising expense and other outlays of promotion, it was stipulated, must be borne by Nat C. Goodwin & Co. and none by the mining corporation.

What was the system? How was it done?

A RACE OF GAMBLERS

PRIOR TO THE BIRTH of Rawhide I had for seven years catered to the speculative (gambling) instinct of the American public, chiefly in building mining camps and financing mining enterprises. I now realized that in order to make a success of the undertaking before me, namely, to put the new camp of Rawhide on the investment map, I must again appeal loudly to the country's gambling instinct.

Maybe you think, dear reader, that a man who caters to the gambling instinct of his fellow men, be his intentions honest or dishonest, is a highly immoral person. Is he?

Do you know that the gambling instinct is responsible for the wonderful growth of the mining industry in the United States? Would you believe that without the gambling instinct the development of the great natural resources of this country would be almost impossible?

With rare exceptions every successful mining enterprise in the United States has been financed in the past by appealing directly to the gambling instinct. In the decade antedating this year considerably more than a billion dollars was raised and invested in this way.

Conservative investors who are satisfied with from three to six per cent on their money do not buy mines or mining stocks. Speculators (gamblers) who are willing to risk part of their fortune in the hope of gaining fivefold or more in a year or a few years—these are the kind who invest in mines and mining stocks.

There are legions of these. Not less than 500,000 men and women in the United States, according to the best statistical information obtainable, are stockholders of mining companies.

In fact, the gambling instinct finds employment in the mining industry long before a property has reached the stage where it can be classed as even a prospect worthy of exploration. The prospector who follows his burro into the mountain fastnesses or across the desert wastes often gambles his very life against the success of his search; those who grubstake him gamble their money.

The gambling instinct seems bound to continue to play an important rôle in the mining industry for all time, or until either the fortune hunting instincts of man are eradicated or all the treasures of the world shall have been mined.

Now, if the practice of catering to the gambling instinct is baneful, I'm a malefactor. So, too, would then be such lofty pinnacled financiers as Messrs. Rothschild, Rockefeller, Morgan, the Guggenheims and others. My own thought is that it is *custom* and the times which are responsible for the maintenance of the great game, and not individuals. The truth is, we are a race of gamblers and we allow the captains of industry to deal the game for us.

Next to money and political power, publicity is recognized by all doers as the most powerful lever to accomplish big things. Not infrequently publicity will accomplish what neither money nor political power can. Generally, publicity can be secured and controlled by either money or political power.

When Rawhide was born I had neither money nor political power. The camp needed publicity. I had nothing to secure publicity with but my wit. I promptly requisitioned what wit I had, and used all of it.

There is an important difference between owning a series of excellent gold mine prospects, which have tremendous speculative possibilities, and the public recognizing them to be such. It is one thing for a manufacturer to be himself assured that his article is a better product for the money than that of his competitor. It is another thing for the consumer to be convinced. Therein lies the value of organized publicity.

To focus the attention of the great American investing public on the camp of Rawhide was the proposition before me. How was this to be accomplished? Display advertising in the newspapers is costly and requires large capital; the purchase of reading notices in publications which accept that class of business, even more so.

One major fact stood out from my early experience as a publicity agent in Goldfield. Few news editors have the heart to consign good copy to the wastepaper basket, particularly if it contains nothing which might cause a comeback.

I resolved to "press agent" the camp.

Mining men from Rhyolite came to Rawhide in response to extensive promotion

8. Publicity and the Public's Money

Probably the most scientifically press agented camp in Nevada had been Bullfrog. Bullfrog was born two years after Goldfield. The Goldfield publicity bureau by this time had greatly improved its art and its efficiency.

When the Bullfrog boom was still young the late United States Senator Stewart, an octogenarian and out of a job, traveled from Washington, at the expiration of his term, to the Bullfrog camp. There he hung out his shingle as a practising lawyer. Immediately the press bureau secured a cabinet photo of the venerable lawmaker and composed a story about his fresh start in life on the desert. The yarn appealed so strongly to Sunday editors of the great city dailies throughout the country that Bullfrog secured for nothing scores of pages of priceless advertising in the news columns.

The Senator built a home, the story said, on a spot where, less than a year before, desert wayfarers had died of thirst and coyotes roamed. The interior of the house on the desert was minutely described. Olive-colored chintz curtains protected the bearded patiarch, while at work in his study, from the burning rays of the sun. Old Florentine cabinets, costly Byzantine vases, and matchless specimens of Sèvres, filled his living rooms. Silk Persian rugs an inch thick decked the floors. Venetian-framed miniature paintings of former Presidents of the United States and champions of liberty of bygone days graced the walls. Costly bronzes and marble statuettes were strewn about in profusion. Visitors could not help deducing that the Senator thought nothing too good for his desert habitat. The name of Bullfrog exuded from every paragraph of the story; also the name of a mine at the approach to which this desert mansion was reared and in the exploitation of which the press agent had a selfish interest.

The remarkable part of this tale, which was printed with pictures of the Senator in one metropolitan newspaper of great circulation and prestige to

the extent of a full page on a Sunday and was syndicated by it to a score of others, was that the only truth contained in it happened to be the fact that the Senator had decided to make Bullfrog his home with a view to working up a law practise. But it was a good story from the Bullfrog press agent's standpoint and from that of the Sunday editor, and even the Senator did not blink at it. He recognized it as camp publicity of the highest efficiency, as did other residents of Bullfrog.

During the Manhattan boom, which followed that of Bullfrog, the publicity bureau became more ambitious. It made a drive at the news columns of the metropolitan press on week days, and succeeded.

At that time the Sullivan Trust Co. of Goldfield was promoting the Jumping Jack-Manhattan mining Co. James Hopper, the gifted magazinist, wrote a story in which the names "Jumping Jack" and "Sullivan Trust Co." appeared in almost every other line. It was forwarded by mail to a great daily newspaper of New York and promptly published as news. The yarn told how the man in charge of the gasoline engine at the mouth of the Jumping Jack shaft had gone stark mad while at work and how but for the quick intervention of the president of the Sullivan Trust Co., who happened to be on the ground, a tragedy might well have been the result.

The miner, the story said, stepped into the bucket at the head of the shaft and asked the man in the enginehouse to lower him to a depth of 300 feet. Quick as a flash the bucket was let down. When the 200-foot point was reached there was a sudden stop. With a rattle and a roar the bucket was jerked back to within 50 feet of the surface. Thereupon it was again lowered and quickly raised again, and the operation constantly repeated until the poor miner became unconscious and fell in a jangled mass to the bottom of the bucket.

Hearing the miner's early cries, Mr. Sullivan had gone to the rescue. He knocked senseless the man in the enginehouse and pinioned him. Then he brought up the bucket containing the almost inanimate form of the miner.

Turning to the demon in charge of the engine, who had now recovered consciousness, Mr. Sullivan cried,

"How dare you do a thing like this?"

The man responded, "His name is Jack, ain't it?"

"Well, what of it?" roared Mr. Sullivan.

"Oh, I was just *jumping the jack!*" chuckled the madman.

This nursery tale was conspicuously printed in a highclass New York newspaper's columns as real news. Undoubtedly the reason why the editors allowed it to pass was that it was believed to be true, but above all was cleverly written.

I was too busy during the early part of the Rawhide boom to do any writing of consequence or even to suggest particular subjects for stories. It seemed to me that the exciting events of everyday occurrence during the progress of the mad rush would furnish the correspondents with enough matter to keep the news pot constantly boiling. I assembled around me the shining lights of the Reno newspaper fraternity and put them on the payroll.

For weeks an average of at least one column of exciting Rawhide news was published on the front pages of the big Coast dailies. The publicity campaign went merrily on. I kept close watch on the character of the news that was being sent out and was pleased in contemplating the fact that very little false coloring, if any, was resorted to. A boisterous mining camp stampede, second only in intensity to the Klondike excitement of eleven years before, was in progress, and there was plenty of live news to chronicle almost every day.

After returning from one of my trips to Rawhide I became alarmed on reading on the front page of the leading San Francisco newspapers a harrowing two-column story about the manner in which Ed Hoffman, mine superintendent of the Rawhide Coalition, had been waylaid the day before on a dark desert road and robbed of $10,000 in gold which he was carrying to the mines for the purpose of discharging the payroll.

I had just left Mr. Hoffman in Rawhide and he had not been waylaid.

I sent for the man who was responsible for the story.

"Say, Jim," I said, "you're crazy. There is a comeback to that yarn that will cost you your job as correspondent for your San Francisco paper. It is rough work. Cut it out!"

"Gee whillikins!" he replied. "How can I? Here's an order for a two column follow-up and I have already filed it."

"What did you say in your second story?" I inquired.

"Well, I told how a posse, armed to the teeth, were chasing the robbers and explained that they're within three miles of Walker Lake in hot pursuit."

"You're a madman!" I protested. "Kill those robbers and be quick; do it tonight so that you choke off the demand for more copy, or you're a goner!"

Next day the correspondent wired to his string of newspapers that the posse had chased the robbers into Walker Lake, where they were drowned.

At the point in Walker Lake where the correspondent said the robbers had found a watery grave it was known to some Reno people that for three miles in both directions the lake was shallow and that the deepest water in that vicinity was less than four feet. This caused some snickering in Reno. Still there was no comeback. The newspapers never learned of the deception.

Walter "Death Valley" Scott was a flamboyant prospector in the Death Valley country; in contrast, George Graham Rice promoted securities and large mining companies around Goldfield and Rawhide.

WALTER SCOTT—"MYSTERIOUS SCOTTY"

Walter Scott, the man with the slouch hat, blue flannel shirt, red necktie, and usually with a wad of greenbacks, has a ranch in upper Death Valley. He is said to have a rich mine in the vicinity. Frequently he has flashed big bankrolls; once chartered a Santa Fe train and made a record run from Los Angeles to Chicago; frequently displayed unusual extravagance by spending large sums of money foolishly; occasionally is seen at Rhyolite, Daggett and other points, going to and from the valley with burros, horses, or his famous team of mules. No one claims to have seen his mine, although it is said he takes out a few thousand now and then to keep him going. He has been pictured in papers and magazines as a miner; maybe so, but his intimate friends have never accused him of swinging a doublejack. At any rate, he gets the money, and the world at large is none the wiser.

The correspondent had been canny enough in sending the story to keep the local correspondents of all other out of town newspapers thoroughly informed. They had sent out practically the same story, and therefore did not give the snap away.

In the early days of the Rawhide boom a rumor reached the camp that Death Valley "Scotty," the illustrious personage who had been press agented from one end of the land to the other as the owner of a secret Golconda, was about to start a stampede into some new diggings. The news bureau decided to kill off opposition. Newspapers of the land were queried as follows:

"Scotty's lair discovered in Death Valley. It is a cache containing a number of empty Wells-Fargo money chests. Scotty has apparently been looting the loot of old-time stage robbers. How many words?"

The newspapers just ate this one up. Column upon column was telegraphed from Nevada. The source of Scotty's wealth being cleared up to the satisfaction of readers of the "yellow," Scotty's value as a mine promoter became seriously impaired.

When I chided the Reno correspondent for sending out the fake story regarding the robbery of Rawhide Coalition's mine manager, I recall that he argued he had made a blunder in one direction only. He said he should have seen to it that the mine manager was actually robbed! That, he said, would have eliminated the danger of a comeback.

Years ago in New York the public was startled by reading of an actress taking her bath in pure milk. A few weeks later newspaper readers were convulsed by stories of another star in the theatrical firmament performing her morning ablutions in a tub of champagne.

"If you don't believe it," said the lady press agent to a lady newspaper reporter sent to cover the story, "I will give you a chance to see the lady in the act."

This was done and, of course, the newspapers were convinced that it was no idle press agent's dream. Of course, neither of these women had been in the habit of bathing in milk or in champagne. A tub of milk costs less than $10 and a tub of champagne less than $200, but you could not have bought this kind of publicity for these performers at anything like such absurdly low figures if you used the display advertising columns of the newspapers. Nor would the advertising have been nearly so effective. The absurd milk story scored a knockout with newspaper readers and earned a great fortune for the actress.

PUBLICITY VIA ELINOR GLYN

AT THIS EARLY STAGE in Rawhide's history the reigning literary sensation of two continents was *Three Weeks*. Nothing, reasoned the correspondents, would attract more attention to the camp than having Mrs. Elinor Glyn at Rawhide, particularly if she would conduct herself while there in a manner that might challenge the criticism of church members.

Sam Newhouse, the multi-millionaire mining operator of Utah, famous on two continents as a charming host, especially when celebrities are his guests, was stopping at the Fairmount Hotel in San Francisco. Mrs. Glyn was in San Francisco at the same time. Mr. Newhouse and Ray Baker, a Reno Beau Brummel, clubman, chum of M. H. De Young, owner and editor of the San Francisco *Chronicle*, and scion of a house that represents the aristocracy of Nevada, were showing Coast hospitality to the distinguished authoress.

A message was sent to Mr. Baker reading substantially as follows: "Please suggest to Mr. Newhouse and Mrs. Glyn the advisability of visiting Rawhide. The lady can get much local color for a new book. If you bag the game, you will be a hero."

Ray was on to his job. Within three days Mrs. Glyn, under escort of Messrs. Newhouse and Baker, arrived in Rawhide after a thirty eight hour journey by railroad and auto from San Francisco.

The party having arrived in camp at dusk, it was suggested that they go to a gambling house and see a real game of stud poker as played on the desert.

They entered a room. Six players were seated around a table. The men were coatless and grimy. Their unshaven mugs, rough as nutmeg graters, were twisted into strange grimaces. All of them appeared the worse for liquor. Before each man was piled a mound of ivory chips of various hues, and alongside rested a six shooter. From the rear trousers' pocket of every player another gun protruded. Each man wore a belt filled with cartridges. Although an impromptu sort of game, it was well staged.

A man with bloodshot eyes shuffled and riffled the cards. Then he dealt a hand to each.

"Bet you $10,000," loudly declared the first player.

"Call that and go you $15,000 better," shouted the second as he pushed a stack of yellows toward the center.

"Raise you!" cried two others, almost in unison.

Before the jackpot was played out $300,000 (in chips) had found its way to the center of the table and four men were standing up in their seats in a frenzy of bravado with the muzzles of their guns viciously pointed at one another. There was enough of the lurking devil in the eyes of the belligerents to give the onlookers a nervous shiver.

Elinor Glyn And Sam Newhouse Stampede Rawhide

ELINOR GLYN ESCORTS MR. NEWHOUSE THROUGH RAWHIDE'S TENDERLOIN

Not in the Roles of Sir and Lady Pandarus but to See a Mining Camp in All Its Varying Hues

"JACK" FLYNN PILOTS THE WAY

Mr. Newhouse Is Impressed with Camp's Showings and Says the Camp Is Bound to Make Good-- Is Examining Mines at Depth Today

By JOSEPH S. JORDAN

(Special Telegram to the Nevada Mining News)

RAWHIDE, May 27.—Rawhide has had many important visitors, men of means, looking for investment, moneyed men in pursuit of speculation, experts to determine what there is in the ground, critics and enthusiasts. All these have had their share of attention, but it remained for the coming of Elinor Glyn, the authoress of "Three Weeks," and Samuel Newhouse and their party to create a profound sensation in this camp of sensations. A year ago, in London, Mr. Newhouse had promised Mrs. Glyn that he would take her to the greatest mining camp in the world when she came to america. Rawhide was unknown then, but when it came to the promise, Rawhide filled the idea that the

When the gun play started, Mrs. Glyn and Messrs. Newhouse and Baker took to the "tall and uncut."

As the door closed and the vanishing forms of the visitors could be seen disappearing around the opposite street corner, all of the men in the room pointed their guns heavenward and shot at the ceiling, which was of canvas. The sharp report of the revolver shots rang through the air. This was followed by hollow groans, calculated to freeze the blood of the retreating party, and by a scraping and scuffling sound that conveyed to the imagination a violent struggle between several persons.

Fifteen minutes later two stretchers, carrying the "dead," were taken to the undertaker's shop. Mrs. Glyn and Mr. Newhouse, with drooped chins, stood by and witnessed the dismal spectacle.

Of course, the "murder" of these two gamblers, during the progress of a card game for sensationally high stakes and in the presence of the authoress of *Three Weeks* made fine front page newspaper copy. Rawhide suggested itself in every paragraph of the stories as a mining center that was large enough to attract the attention of a multi-millionaire mine magnate of the caliber of Sam Newhouse and of an authoress of such worldwide repute as Elinor Glyn. The camp got yards of free publicity that was calculated to convince the public it was no flash in the pan, which was exactly what was wanted.

The next night Elinor Glyn, having recovered from the shock of the exciting poker game, was escorted through Stingaree Gulch. The lane was lined on both sides with dance halls and brothels for a distance of two thousand feet. Mrs. Glyn "sight saw" all of these.

Rawhide scribes saw a chance here for some fine writing:

> The wasted cheeks and wasted forms of frail humanity, as seen last night in the jaundiced light that was reflected by the crimson shaded lamps and curtains of Stingaree Gulch, visibly affected the gifted English authoress. They carried to Mrs. Glyn an affirmative answer to the question, so often propounded recently, whether it is against public morality to make a heroine in *Three Weeks* of a pleasure-palled victim of the upper set. It was made plain to Mrs. Glyn that her heroine differed from the Stingaree Gulch kind only in that her cheeks were less faded than her character.

That's the kind of Laura Jean Libby comment on Mrs. Glyn's tour of Stingaree Gulch that one of the Rawhide correspondents wired to a "yellow,"

Rawhide's famous Stingaree Gulch

with a view to pleasing the editor and to insuring positive acceptance of his copy.

Later in the night a fire alarm was rung in. The local fire department responded in wild western fashion. The conflagration, which was started for Mrs. Glyn's sole benefit, advanced with the rapidity of a tidal wave. It brought to the scene a mixed throng of the riff-raff of the camp. The tumult of voices rose loud and clear. The fire embraced all of the deserted shacks and waste lumber at the foothills of one of the mines. The liberal use of kerosene and a favoring wind caused a fierce blaze. It spouted showers of sparks into the darkness and gleamed like a beacon to desert wayfarers. The fierce yells of the firemen rang far and wide. Of a sudden a wild-haired individual thrust himself out of the crowd and sprang through the door of a blazing shack. He disappeared within the flames. Three feet past the door was a secret passage leading to a shelter in the tunnel of an adjoining mine. Mrs. Glyn, of course, did not know this. She acclaimed the act as one of daring heroism.

Water in the camp was scarce, so there was a resort to barreled beer and dynamite. Soon the flames of the devouring fire were extinguished. Again the newspapers throughout the land contained stories, which were telegraphed from the spot, regarding the remarkable experiences of the much-discussed authoress of "Three Weeks" in the new, great gold camp of Rawhide. The press agent was in his glory.

"AL" MILLER'S SIEGE

ELINOR GLYN'S EXPERIENCES in Rawhide were by no means the most interesting that newspaper readers of the United States were privileged to read during the course of the press agenting of the camp.

"Al" Miller was one of the first experienced mining operators to get into Rawhide. He landed in camp in the early part of 1907. After a thorough inspection of the mine showings throughout the district, he hit upon the Hooligan Hill section of the Rawhide Coalition property as a likely looking spot to develop pay ore. Mr. Miller had been mining for a great many years and had been identified with some important mining projects in Colorado. When he applied for a lease on that section of the Coalition property embracing a good part of Hooligan Hill it was granted to him without parley.

Mr. Miller financed his project right in the camp of Rawhide. He interested five other mining men. A syndicate was formed. Each of these six took an equal interest. All agreed to subscribe to a treasury fund to meet the expense of development.

A shaft was started on a very rich stringer of gold ore. When it had reached a depth of about 40 feet the Miller lease was regarded as one of the big comers of the camp. In fact, a good grade of ore was exposed on all sides and in the bottom of a 4½x7½ foot shaft. Specimens assayed as high as $2,000 a ton.

At this stage of the enterprise an operating company was formed. Those who had formed the original syndicate divided the ownership stock among themselves. Mr. Miller was given full charge and allowed a salary for his services. Day after day you could see him on the job, sharpening steel, turning a windlass to hoist the muck from the bottom of the shaft after each round of shots had been fired, and making a full hand as mine manager, blacksmith, mucker and shift boss.

One day I was sitting in my office at Reno when I received a telephone message that there was a big fight on over the control of the Miller lease. Mr. Miller and a big Swede who was working for him had barricaded themselves at the mine. They threatened death to any one who approached. We had, for a day or two, been hungering and thirsting for some live news of the camp. My journalistic instinct got busy. I queried our Rawhide correspondent. He advised that the situation really looked serious and that a genuine scrap threatened. Mr. Miller had installed a good sized arsenal at the mine and laid in about three days' provisions. He declared that he was prepared to hold out for an indefinite period.

I wired our correspondent at Rawhide instructions to file a story up to 1,000 or 1,500 words. Naturally excitement ran high in the camp. Soon hundreds of people gathered at points of vantage along the crest of Hooligan Hill and surrounding uplifts. Every one was expectantly awaiting interesting developments. To the casual onlooker it seemed as though possibly a score or more who stood ready to storm the mine might become involved. In fact, no one could tell how soon hostilities would break loose.

Using the telegrams I had received from camp, one of my men dictated a story containing the facts and sent it over to the Reno correspondent of the Associated Press. It was put on the wire without a moment's hesitation.

Mr. Miller had formed a rampart about the collar of the shaft. Sacked ore was piled up to a height of about five feet. The gold-laden stuff surrounded the shaft on all sides but one, the exception being to the northwest. There Hooligan Hill slanted upward at an angle of less than twenty degrees from vertical. It was from this approach that Mr. Miller was forced to guard constantly against attack. He found it necessary, according to our dispatches, to keep a constant vigil in order to preclude the possibility of a surprise. He and his Swede companion alternated in keeping the lookout. Occasionally the fitful coughing of the gasoline engine exhausts from the mining plants on Balloon Hill and Grutt Hill were interspersed by the sharp report of a six shooter as the besieged parties either actually or mythically observed a threatened approach of the enemy.

Although the principals cast in this little mimic war were limited to perhaps less than a score, every incident or detail was provided to make up a very threatening and keenly interesting situation, with several lives hanging in the balance. There is no doubt that Mr. Miller at least, and perhaps his Swede companion, would have resisted any attempt to take "Fort Miller," as we styled it, even to sacrificing his life, for he was known as a man of action who had been in numerous critical situations without showing the slightest exercise of the primal instinct. The fact that Rawhide was saved from an episode that might have measured up to the tragic importance of a pitched battle and caused the loss of a number of lives was undoubtedly due to the patient willingness of Mr. Miller's partners and their supporters to satisfy themselves with a siege and to starve out the two men in possession of the mine rather than undertake to rout them.

The story went like wildfire and we were besieged for others and for a follow-up on the original story. For three days we kept the yarn alive and the wires burdened with details of the siege and unsuccessful storming of Camp Miller, Hooligan Hill, Nevada.

I venture to say that Mr. Hearst, with his well known facility for serving up hot stuff to a sensation-loving following, never surpassed in this particular the stories that were scattered broadcast over the United States foundered upon this interesting episode in the mining development of Rawhide.

The story promised to be good for at least a week when we were somewhat surprised to hear that Mr. Miller had capitulated. It seems that in storing his fort with provender he had supplied only one gallon of whisky and when this ran low, on the second or third day, he attempted, single-handed, a foraging expedition in search of a further supply of John Barleycorn. During his absence his Swede campanion hoisted a flag of truce, and when Miller returned to the scene of action he found his mine in the possession of his enemies.

Charles G. Gates, son of John W. Gates, the noted stock market plunger, visited Rawhide twice. He spent his time by day inspecting the numerous mine workings, of which there were not less than seventy five in full blast. At night he was a frequent winner at the gambling tables. His advent in Rawhide was telegraphed far and wide and contributed to excite the general interest.

A young woman of dazzling beauty and fine presence was discovered in camp unchaperoned. She had been attracted to the scene by stories of fortunes made in a night. Under a grilling process of questioning by a few leading citizens she divulged the fact that she had run away from her home in Utah to seek single-handed her fortune on the desert. In roguish manner she expressed the opinion that if allowed to go her own way she would soon succeed in her mission. But she would not divulge the manner in which she proposed to operate. She confessed she had no money. There was a serene but settled expression of melancholy in her eyes that captivated everybody who saw her.

Many roving adventurers of the better class in the district who had listened to the call of the wild yet would have felt as much at home in the salon of a Fifth Avenue millionaire as in the boom camp, pronounced her beauty to be in a class by itself. There was no law in the camp which would warrant the girl's deportation, yet action appeared warranted. Within a few moments $500 was subscribed as a purse to furnish the girl a passage out of camp and for a fresh start in life. The late Riley Grannan, race track plunger, Nat C. Goodwin, the noted player, and three others subscribed $100 each. She refused to accept the present. Next day she disappeared.

There was a corking human interest story here. Newspapers far and wide published the tale. Two years later this girl's photograph was sent without her knowledge to the judges of a famous beauty contest in a far western state. The judges were on the point of voting her the prize without question when investigation of her antecedents revealed her Rawhide escapade. The award was given to another.

When the camp was four months old and water still commanded from $3 to $4 a barrel, the standard price for a bath being $5, a banquet costing $50 a plate was served to one hundred soldiers of fortune who had been drawn to the spot from nearly every clime. The banqueters to a man played a good knife and fork. The spirit of *camaraderie* permeated the feast. There was much libation, much postprandial speechifying, much unbridled joyousness. *Bon mots* flew from lip to lip. Song and jest were exchanged. The air rang with hilarity. Nat C. Goodwin warmed up to a witty, odd, racy vein of across-the-table conversation. Then he made a felicitous speech. Others followed him in similar vein. Luxuriant and unrestrained imagination and slashiness of wit marked most of the talks. The festivities ended in a revel.

The correspondents burned up the wires on the subject of that banquet. In the memory of the most ancient prospector no scene like this had ever been enacted in a desert mining camp when it was so young and at a time

RILEY GRANNAN'S FUNERAL

An Oration by W. H. Knickerbocker *at* Rawhide, Nevada, April 3rd, 1908

George Graham Rice once said, "When all the gold in Rawhide's towering hills shall have been reduced to bullion and not even a post is left . . . posterity will remember Rawhide for the funeral oration that was pronounced over the bier of Riley Grannan by W. H. Knickerbocker."

The scene was dramatic and impressive. Around the bier in the rear of the saloon gathered rough, unkempt miners in high boots, just from their work. There were also gamblers and floaters and girls from the dance halls, all gathered together to pay respect to Riley Grannan, a fellow gambler and adventurer. Veteran prospector W. H. Knickerbocker stood near the rude coffin and said:

❦

I feel that it is incumbent upon me to state that in standing here I occupy no ministerial or prelatic position. I am simply a prospector. Riley Grannan was born in Paris, Kentucky. He died day before yesterday in Rawhide.

This is a very brief statement. You have the birth and the period of the grave. Who can fill the interim? Who can speak of his hopes and fears? Who can solve the mystery of his quiet hours that only himself knew? I cannot. Here is the beginning and the end. He was born in the Sunny Southland, where brooks and rivers run musically through the luxuriant soil. He died in Rawhide, where in winter the shoulders of the mountains are wrapped in garments of ice and in summer the blistering rays of the sun beat down upon the skeleton ribs of the desert. Is this a picture of universal human life?

Sometimes when I look over the circumstances of human life, a curse rises to my lips, and, if you will allow me, I will say here that I speak from an individual point of view. I cannot express other than my own views. If I run counter to yours, at least give me credit for a desire to be honest.

When I see the ambitions of a man defeated; when I see him struggling with mind and body in the only legitimate prayer he can make to accomplish some end; when I see his aim and purpose frustrated by a fortuitous combination of circumstances over which he has no control; when I see the outstretched hand, just about to grasp the flag of victory, take instead the emblem of defeat, I ask: What is life? Dreams, awakening and death; "a pendulum 'twixt a smile and a tear"; "a momentary halt within the waste and then the nothing we set out from"; "a walking shadow, a poor player that struts and frets his hour upon the stage and then is heard no more"; "a tale told by an idiot, full of sound and fury, signifying nothing"; a child-blown bubble that but reflects the light and shadow of its environment and is gone; a mockery, a sham, a lie, a fool's vision; its happiness but Dead Sea apples; its pain the crunching of a tyrant's heel.

If I have gauged Riley Grannan's character correctly, he accepted the circumstances surrounding him as the mystic officials to whom the universe had delegated its whole office concerning him. He seemed to accept both defeat and victory with equanimity. He was a man whose exterior was as placid and gentle as I have ever seen, and yet when we look back over his meteoric past we can readily understand, if this statement be true, that he was absolutely invincible in spirit. If you will allow me, I will use a phrase most of you are acquainted with. He was a "dead game sport." I say it not irreverently, but fill the phrase as full of practical human philosophy as it will hold, and I believe that when you say one is a "dead game sport" you have reached the climax of human philosophy.

I know that there are those who will condemn him. There are those who believe today that he is reaping the reward of a mis-spent life. There are those who are dominated by mediaeval creeds. To those I have no words to say in regard to him. They are ruled by the skeleton hand of the past and fail to see the moral beauty of a character lived outside their puritanical ideas. His goodness was not of that type, but of the type that finds expression in a word of cheer to a discouraged brother; the type that finds expression in quiet deeds of charity; the type that finds expression in friendship, the sweetest flower that blooms along the dusty highway of life; the type that finds expression in manhood.

He lived in the world of sport. I do not mince my words. I am telling what I believe to be true. In the world of sport — hilarity sometimes, and maybe worse—he left the impress of his character on this world, and through the medium of his financial power he was able with his money to brighten the lives of its inhabitants. He wasted it so the world says. But did it ever occur to you that the most sinful men and women who live in this world are still men and women? Did it ever occur to you that the men and women who inhabit the night-world are still men and women? A little happiness brought into the lives of the straight and the good. If you can take one ray of sunlight into their night-life and thereby bring them one single hour of happiness, I believe you are a benefactor.

Riley Grannan may have "wasted" some of his money this way.

We stand at last in the presence of the Great Mystery. I know nothing about it, nor do you. We may have our hopes, but no knowledge. I do not know whether there be a future life or not; I do not say there is not. I simply say I do not know. I have watched the wicket-gate closed behind many and many a pilgrim. No word has come back to me. The gate is closed. Across the chasm is the gloomy cloud of death. I say I do not know. And, if you will allow this expression, I do not know whether it is best that my dust or his at last should go to feed the roots of the grasses, the sagebrush or the flowers, to be blown in protean forms by the law of the persistency of force, or whether it is best that I continue in personal identity beyond what we call death. If this be all, "after life's fitful fever, he sleeps well; . . . Nothing can harm him further." God knows what is best.

This may be infidelity, but if it is, I would like to know what faith means. I came into this universe without my volition—came and found a loving mother's arms to receive me. I had nothing to do with the preparation for my reception here. I have no power to change the environment of the future, but the same power which prepared the loving arms of a mother to receive me here will make proper reception for me there. God knows better than I what is good for me, and I leave it with God.

As we stand in the presence of death, we have no knowledge, but always, no matter how dark the gloomy clouds hang before me, there gleams the star of hope. Let us hope, then, that it may be the morning star of eternal day. It is dawning somewhere all the time. Did you ever pause to think that this old world of ours is constantly swinging into the dawn? Down the grooves of time, flung by the hand of God, with every revolution it is dawning somewhere all the time. Let this be an illustration of our hope. Let us believe, then, that in the development of the human soul, as it swings forward toward its destiny, it is constantly swinging nearer and nearer to the sun.

And now the time has come to say good-bye. The word "farewell" is the saddest in our language. And yet there are sentiments sometimes that refuse to be confined in that word. I will say "Good-bye, old man." We will try to exemplify the spirit manifested in your life in bearing the grief at our parting. Words fail me here. Let these flowers, Riley, with their petaled lips and perfumed breath, speak in beauty and fragrance the sentiments that are too tender for words. Good-bye.

when the country was just emerging from a panic that seemed for a while to warp its whole financial fabric.

THE FUNERAL ORATION FOR RILEY GRANNAN

IN APRIL 1908, Riley Grannan, the noted race track plunger, died of pneumonia in Rawhide, where he was conducting a gambling house. He was ill only a few days and his life went out like the snuff of a candle. When all the gold in Rawhide's towering hills shall have been reduced to bullion and not even a post is left to guide the desert wayfarer to the spot where was witnessed the greatest stampede in Western mining history, posterity will remember Rawhide for the funeral oration that was pronounced over the bier of Mr. Grannan by H. W. Knickerbocker, wearer of the cloth and mine promoter.

The oration delivered by Mr. Knickerbocker on this occasion was a remarkable example of sustained eloquence. Pouring out utterances of exquisite thought and brilliant language in utter disregard of the length of his sentences and without using so much as a pencil memorandum, Mr. Knickerbocker with a delicacy of expression pure as poetry urged upon his auditors that the deceased dead game sport had not lived his life in vain. Soon the crowd, who listened with rapt attention, was in the melting mood. As Mr. Knickerbocker progressed with his discourse his periods were punctuated with convulsive bursts of sorrow.

Rawhide correspondents reorganized the full value of the occasion from the press agent's standpoint. Mr. Grannan had been a world famous plunger on the turf, and the correspondents burned the midnight oil in an effort to do their subject justice.

Some other lights and shadows of Rawhide press agenting are contained in the following dispatch, which appeared in a San Francisco newspaper in the early period of the boom:

> GOLDFIELD, February 19.—W. H. Scott of the Goldfield brokerage house of Scott & Amann, who returned from Rawhide this morning, expresses the opinion that within a year that camp will be the largest gold producer in the State. "When a man is broke in Rawhide," said Mr. Scott, "he can always eat. All he has to do is to go to some lease and pan out breakfast money. There is rich ore on every dump, and every man is made welcome."

H. W. Knickerbocker sent this one to a Reno newspaper:

> Gold! Gold! Gold! The wise men of old sought an alchemy whereby they could transmute the base metals into gold. It was a fruitless quest then; it is a needless quest now. Rawhide has been discovered! No flowers bloom upon her rock-ribbed bosom. No dimpling streams kiss her soil into verdure, to flash in laminated silver 'neath the sunbeam's touch. No flowers nor food, no beauty nor utility on the surface; but from her desert covered heart Rawhide is pouring a stream of yellow gold out upon the world which is translatable, not simply into food and houses and comfort, but also into pictures and poetry and music and all those things that minister in an objective way to the development of a full-orbed manhood.

Joseph S. Jordan, the well known Nevada mining editor, filed this dispatch to the newspapers of his string on the Coast:

> Right through what is now the main street of Rawhide, in the days of '49, the makers of California passed on their way to the new Eldorado. They had many hardships through which to pass before reaching the gold which was their lure, and thousands that went through the hills of Rawhide never reached their goal. They were massacred by the Indians, or fell victims to the thirst and heat of the desert, and for many years the way across the plains was marked by the whitening bones of the pathfinders. And here all the while lay the treasures of Captain Kidd, the ransoms of crowns.

Harry Hedrick, the veteran journalist of far western mining camps, sent his newspapers this:

> To stand on twenty different claims in one day, as I have done; to take the virgin rock from the ledge, to reduce it to pulp and then to watch a string of the saint-seducing dross encircle the pan; to peer over the shoulder of the assayer while he takes the precious button from the crucible—these are the convincing things about this newest and greatest of gold camps. It is not a novelty to have assays run into the thousands. In fact, it is commonplace. To report strikes of a few hundred dollars to the ton seems like an anticlimax.

There were scores of actual happenings in Rawhide that make it possible for me to say in reviewing the vigorous publicity campaign which marked its first year's phenomenal growth, that ninety per cent of the correspondence, including the special dispatches sent from the camp and from Reno, which was published in newspapers of the United States, was not only based on fact but was literally true in so far as any newspaper reporter can be depended upon accurately to describe events.

Ask any high class newspaper owner or editor to express his sentiments regarding the faking which formed about ten per cent of the Rawhide press work described herein and he will tell you that such work is a reproach to journalism. Maybe it is, but we are living in times when such work on the part of press agents is the rule and not the exception. The publicity agent who can successfully perform this way is generally able to command an annual stipend as big as that of the President of the United States. There was nothing criminal about the performance in Rawhide, because there was no intentional misrepresentation regarding the character or quality of any mine in the Rawhide camp. Correspondents were repeatedly warned to be extremely careful not to overstep the bounds in this regard.

Confessedly there are grades of faking which no press agent would care to stoop to.

Somewhere in De Quincey's "Confessions of an Opium Eater" he describes one of his pipe dreams as perfect moonshine, and, like the sculptured imagery of the pendulous lamp in "Christabel," *all carved from the carver's brain*. Rawhide and Reno correspondents were guilty of very little work which De Quincey's description would exactly fit. There was a basis for nearly everything they wrote about, even the alleged discovery of Death Valley Scotty's secret storehouse of wealth, that story having been in circulation in Nevada, although not theretofore published, for upward of eighteen months. Unsubstantial, baseless, ungrounded fiction had been resorted to, it is true, during the Manhattan boom, in a single story about the madman in charge of the hoist on the Jumping Jack, but this was an exception to the rule and the story was harmless.

AMONG THE BIG FELLOWS

If you don't think the character of the press agent's work during the Rawhide boom was comparatively high class and harmless, dear reader, you really have another "think" coming. At a time when Goldfield Consolidated was wobbling in price on the New York Curb and the market needed support, just prior to the smash in the market price of the stock from $7 to around $3.50, the New York *Times* printed in a conspicuous position on its financial page a news story to the effect that J. P. Morgan & Co. were about to take over the control of that company. That's an example of a *harmful* fake, the coarse kind that Wall Street occasionally uses to catch suckers.

Here is another: Thompson, Towle & Co., members of the New York Stock Exchange, issue a weekly newspaper called the *News Letter*. Much of its space is given over to a review of the copper situation, at the mines and in the share markets. W. B. Thompson, head of the firm, he of Nipissing market manipulation fame, is interested to the extent of millions in Inspiration, Utah Copper, Nevada Consolidated, Mason Valley and other copper mining companies. On January 25, 1911, when both the copper metal and copper share markets were sick, and both the price of the metal and the shares were on the eve of a decline, which temporarily ensued, the *News Letter*, in an article headed "Copper," said that "every outcrop in the country has been examined and it is not known where one can look for new properties."

The readers of the *News Letter* were asked to believe that no more copper mines would be discovered in this country and that, because of this and other conditions which it mentioned, the supply of the metal must soon be exhausted and the price of the metal and of copper securities must advance.

The statement in the *News Letter* that every outcrop in the country has been examined and that it is not known where one can look for new properties —well, if the whole population of North America agreed in a body to accept the job of prospecting the Rocky Mountains and Sierra Nevada Mountains alone they could hardly perform the job in a lifetime.

The use of the automobile has undoubtedly been responsible in the past few years for an impetus to the discovery of mines which is calculated to double the mineral product of this country in the next two decades, and who shall say what the flying machine will accomplish in this regard? Further, new smelting processes and improved reduction facilities generally are daily reducing the cost of treatment of ores and are making commercially valuable low-grade ore bodies heretofore passed up as worthless.

The best opinion of mining men in this country is that our mineral resources have not yet been skimmed and that the mining ground of the Western country has not yet been well scratched.

Therefore, the statement made in a newspaper which is supposedly devoted to the interests of investors, that it need not be expected that more copper mines are going to be discovered, is a snare calculated to trap the unwary.

The foregoing is an example of a very harmful but comparatively crude fake, employed by some promoters in Wall Street of the multi-millionaire class when their stocks need market support.

Here is a specimen of the *insidious* brand of get-rich-quick fake. On March 7, 1911, the New York *Sun* printed in the second column of its front page the following dispatch:

> TACOMA, WASH., March 6.—F. Augustus Heinze has struck it rich again; this time it's a fortune in the Porcupine gold fields in Canada.
>
> Charles E. Herron, a Nome mining man, who has just returned from the new gold fields, is authority for the statement that Heinze is inside the big money. He has bought the Foster group of claims, adjoining the celebrated Dome mine, from which it is estimated that $25,000,000 will be gleaned this year and for the development of which a railroad is now under construction.
>
> The Porcupine gold field, according to Herron, is one of the wonders of the age. One prospector has stripped the vein for a distance of fifty feet and polished it in places, so that gold is visible all along. His trench is three feet deep and he asks $200,000 cash for it as it stands.
>
> A party of Alaskans offered the owner of this claim $50,000 a shot for all the ore that could be blown out with two sticks of dynamite, but he refused.

Press work like the foregoing is more than likely to separate the public wrongfully from its money.

The item serves as an excellent example of one of "the impalpable and cunningly devised tricks that fool the wisest and which landed you" that I promised, at the beginning of "My Adventures with Your Money," to lay bare. I said in my foreword:

> Are you aware that in catering to your instinct to gamble, methods to get you to part with your money are so artfully and deftly applied by the highest powers that they deceive you completely? Could you imagine it to be a fact that in nearly all cases where you find you are ready to embark on a given speculation, ways and means that are almost scientific in their insidiousness have been used upon you?

The New York *Sun* article says it is estimated that $25,000,000 will be gleaned this year from the Dome mine in Porcupine. The truth is, no engineer has ever appraised the ore in sight in the entire mine, according to any statements yet issued, at anything like half of that amount gross, and the mine itself can not possibly produce so much as $100,000 this year.

A mill of 240 tons per diem capacity has been ordered by the management and it is expected will be in operation by October first, but no sooner. The ore, according to H. P. Davis's *Porcupine Hand Book*, an accepted authority, "has been stated to average from $10 to $12 a ton." The lowest estimated cost of mining and milling is $6. A fair estimate of profits would, therefore, be $5 per ton, not allowing for any expenses of mine exploration in other directions on the property or other incidental outlay, which will undoubtedly amount to $1 per ton on the production. The production of 240 tons of ore per day at $4 per ton net profit would net returns of $28,800 per month. If the mill runs throughout October, November and December of this the company will glean $86,400 during 1911, and not $25,000,000, as the New York *Sun* article suggests.

In arriving at these figures I am more than fair. Recent estimates of the average value of the ores is $8, and I know of some estimates by very competent mining men that are as low as $4. Some engineers say justification is lacking for even a $4 estimate. The Dome is by no means a proved commercial success as yet from the mine standpoint, although the possessor of much ore, because of the uncertain average values.

How great an exaggeration the New York *Sun's* $25,000,000 estimate is may be gathered from the statement that to glean $25,000,000 in one year from any mine where the ore assays $11 on an average, and the cost of mining, milling and new development is $7, the gross value of the tonnage in the mine that is milled during the one year must be at least $53,571,000. Further, to reduce such a quantity of that quality of ore to bullion in a single year would require the erection of mills of 17,260 tons per day capacity. As mentioned, the actual per diem capacity of the mill now under construction is 240 tons.

Undoubtedly the Dome mining company flotation will soon be made and the public will be allowed to subscribe for the shares or buy them on the New York Curb at a figure agreeable to the promoters. This seems certain, for otherwise why this raw press work? [The foregoing comment on the Porcupine situation has been more than justified by developments after the date of this writing. The first battery of forty stamps in the first stamp mill was not in operation till April 1912, more than a year from the date of the prediction that $25,000,000 would be gleaned in 1911.]

The article says that a number of Alaskans offered money at the rate of $50,000 a shot for all the ore that could be blown out with two sticks of dynamite, but were refused. There never was a statement made by any wildcatter now behind prison bars in any literature I ever saw that could approach this one in flagrant misrepresentation of facts. All the ore that could be displaced in one shot with two sticks of dynamite would not exceed four tons. In order to repay the investor it would be necessary, therefore, that this ore average better than $12,500 per ton. The New York *Sun's* story says that notwithstanding this offer the owner was willing to sell the whole property for $200,000. Imagine this: there are four tons of rock on the property worth $12,500 per ton, for a distance of 50 feet the gold shimmers on the surface, and there are hundreds of thousands of tons of rock in the same kind of formation on the same property, but still the owner is willing to dispose of all of it for $200,000! The statement is preposterous and outrageous. It is the kind described by De Quincey as "all carved from the carver's brain."

THE REVERSE ENGLISH

Now, ABOUT THE reverse English in this line of press work. Similar ways and means, dear reader, that are just as scientific in their insidiousness have been used upon you to poison your mind *against* the value of mining investments of competing promoters, when it has been found to the interest of powerful men to bring this about.

When the offices of B. H. Scheftels & Co., with which I was identified, were raided in seven cities by special agent Scarborough (since permitted to resign) of the Department of Justice of the United States government, in September, 1910, two of the men who had been active in bringing about the raid assembled in the parlor of the Astor House the newspaper men assigned to cover the story by New York and Brooklyn newspapers. There they gave out the information that Ely Central, which I had advised the purchase of at from 50 cents per share up to $4 and down again, was actually under option to me and my associates in large blocks at five cents.

As a matter of fact, the average price paid over for this option stock in real hard money by my people was in excess of 90 cents per share, without adding a penny to the cost for expenses of mining engineers, publicity or anything else. My people had also partly paid for a block bought at private sale at the rate of $3 a share, besides buying tens of thousands of shares in the open market at $4 and higher. The New York *Times* and the New York *Sun*, two newspapers which make capital of the rectitude of both their news and advertising columns, published this statement, along with forty others that were just as false, if not more so. So did the New York *American* and the other Hearst newspapers of the United States.

The New York *Times* story related how I had personally cleaned up in fifteen months not less than $3,000,000 as the result of my market operations.

As a matter of fact, I and my associates had impoverished ourselves trying to support the stock in the open market against the concerted attacks of rival promoters and other powerful interests on whose financial corns we had tread. Every well informed person in Wall Street knows this.

The New York *Times* stated that every man connected with B. H. Scheftels & Co. had tried to obtain membership on the New York Curb and that all of the requests were turned down. No application was ever made for membership because, first, the rules of the Curb forbade corporation memberships, and, second, the Scheftels company already employed several members on regular salary and more than a dozen members on a commission basis.

It was also stated that B. H. Scheftels & Co. applied to the Boston Curb for membership and that their application was rejected. This was also a lie made out of whole cloth.

In three months, the New York *Times* said, no less than 400,000 letters had been received in reply to circulars sent out by B. H. Scheftels & Co. This is an average of over 5,000 letters for each business day during the period of three months. The exaggeration here was about 5,000 per cent.

All of the properties promoted by the Scheftels company were stated in the New York *Times* article to be practically worthless. This was utter rubbish and so misleading that had I been accused of pocketpicking the effect could not have been more harmful.

Rawhide Coalition had produced upward of $400,000 in gold bullion, had probably been high-graded to the extent of that much more, according to the judgment of well posted men on the ground, and not less than five miles of underground development work had been done on the property. Development work and production had never ceased for a day. Besides, when the Rawhide camp was still in its swaddling clothes, I had originally purchased the controlling interest for Nat C. Goodwin & Co. at a valuation of $700,000 for the mine.

The control of Ely Central had been taken over by B. H. Scheftels & Co. and paid for at a valuation well in excess of a million dollars for the property, and upward of $200,000 had been spent in mine development during the fourteen months of the Scheftels quasi-control. Jumbo Extension was a famous producer of Goldfield. Subsequent to the raid one-twentieth of its acreage was sold to the Goldfield Consolidated for $195,000. On July 15th of the current year the company disbursed to stockholders $95,000 in dividends, being 10 per cent on the par of the issued capitalization. Bovard Consolidated, which was promoted at 10 cents a share as a speculation, had turned out to be a lemon after a period of active mine development, the values in the ore pinching out at depth, but B. H. Scheftels & Co. had immediately informed stockholders to this effect.

The New York *Times* stated that B. H. Scheftels & Co. sold Ely Central stock to the amount of five or six millions in cash and made a profit of $3,000,000 on the transaction. The books of the Scheftels company show that the company not only made no money on the sale of Ely Central but actually lost vast sums.

The New York *Times* said that it had been advertised that a carload of ore had been shipped from the Ely Central mine as a sample, but that the government had not been able to find out to whom this carload was consigned. The truth was that the consignment had been made to the best known smelter company in the United States, that the ore averaged seven per cent copper, and that it could not have been shipped out of camp except over a single railroad which has the monopoly—an easy transaction to trace.

B. H. Scheftels & Co. were accused by the New York *Times* of clearing up nearly $600,000 in three months on the promotion of the South Quincy Copper Co. The facts were that, after receiving $30,000 in subscriptions and returning every subscription on demand because of the slump in metallic copper, the Scheftels company abandoned the promotion and never even applied for listing of the stock in any market. A large sum was lost by the Scheftels company here.

Even in stating the penalty for misuse of the mails, which was the crime charged by the government agent who afterwards resigned as a consequence of conduct objectionable to the government, the New York *Times* stated that the punishment was five years in prison, which was more hop-skip and go-merry mistaking. The crime is a misdemeanor and the maximum penalty for an offense is eighteen months.

I have counted not less than five hundred unfounded and misleading statements of this kind regarding myself and associates that have been made in the past year by newspapers and press associations. The shadow has been taken for the substance.

Now, the Scheftels raid, I shall prove in due time, was the culmination of as bitterly waged a campaign of misrepresentation and financial brigandage as has ever been recorded. Chronologically an introduction of the subject is out of place here. The effect, however, of the press agenting which formed a part of the campaign of destruction is pertinent to the topic under consideration.

The immediate result was that thousands of stockholders in the various mining companies that had been sponsored by the Scheftels company were robbed of an aggregate sum amounting into millions, which represented the ensuing decline in market value of the stocks.

The newspaper campaign of misrepresentation and villification was essential to the plans and purposes of the men who sicked the government on to me. The final destruction of public confidence in the securities with which I was identified became necessary to justify the whole proceeding in the public mind.

On the surface of the play it was made to appear that the government of the United States had reached out righteously for the suppression of a dangerous band of criminals. The story in the New York *Times* and other newspapers on the day after the raid was justification made to this end.

The fact that tens of thousands of innocent stockholders might lose their all, as a result of the foul use of powerful maladroit publicity machinery, did not stop the conspirators for a moment. I had a youthful past and, therefore, the newspapers took little chance in publishing anything without investigation and proof that might be offered. And they went the limit, particularly those newspapers that are in the habit of permitting the use of their news columns from time to time to help along the publicity measures of powerful interests.

Contrasted with the comparatively harmless faking that characterized Rawhide's press agenting, the raw work of the newspapers just described is as different as angel cake from antimony. If you are not yet convinced, hearken to this:

THE POWER OF THE PUBLIC PRINT

IN THE *Saturday Evening Post* of December 31, 1910, there appeared an article by Edward Hungerford headed, "Launching a Corporation. How the Pirates and Merchantmen of Commerce Set Sail," from which I quote, without the omission or change of so much as a comma. Referring, in my opinion to Ely Central, promoted by myself and associates, Mr. Hungerford says:

> Here is a typical case—a mining property recently exploited on the curb market, the shipyard of many of these pirate craft: a prospect located not far from one of the bonanza mines of the West was capitalized by a number of men who, after they had convinced themselves that it would not pay, dropped it and gave little thought to the company they had organized.
>
> One day they received through a lawyer an offer of four thousand dollars for the even million shares of stock they had

prepared to issue at a face value of five dollars a share. They were told a wealthy young man was willing to take a four thousand dollar flier on the property, on the outside chance that it might develop ore. The deal was made. Soon after a well known man was named as a part owner of the mine, which promised to enrich all those interested in it.

That was not the first time that the marketable value of a name that is known had been used to exploit a corporation. Any man of standing has many such offers.

The shares of stock that had been purchased for four cents each were peddled on the curb at fifty cents. Then they were advanced to sixty cents. Soon a "market"—so called—was made and the stock found a ready sale. Point by point it was advanced until it actually was eagerly sought by investors, who were not only willing but eager to pay four dollars a share for it.

Mr. Hungerford states in the foregoing: "This mine was capitalized by a number of men who dropped out after they convinced themselves that it would not pay." The statement is false if it refers to Ely Central, as I believe it does. The chief owners and organizers attempted to promote it through a New York Stock Exchange house on the New York Curb at above $7 per share, or at a valuation of more than $8,000,000 for the mine, but the bankers' panic of 1907-08 intervened, and for that reason they quit. The stock sold in 1906 at above $7.50 a share on the New York Curb, two years before I became identified with it.

Mr. Hungerford says that one day these men received through a lawyer an offer of $4,000 for a million shares of stock, and they sold.

How cruelly false this statment is nobody can feel more than myself. The average price paid by my associates in hard money for the controlling interest in the 1,600,000 shares of capitalization, as already mentioned, was above 90 cents, or considerably more than one million dollars in all. An additional $600,000 or more was used to protect the market for the stock, making our cost, without adding a cent for promotion expenses, about $1.50 per share instead of four cents—more than $2,000,000 for the property and not $5,000.

Line by line and word for word I could analyze the statement of Mr. Hungerford and show that 95 per cent of it is false both in premise and deduction. But this would be only cumulative on the one point. My excuse for mentioning the item is to give a striking example of the startling force and power which attaches to insidious newspaper publicity of the kind quoted from the New York *Times*. Mr. Hungerford fell for it, and innocently lent himself to the purposes of the men who sponsored the story by himself passing it on to the readers of the *Saturday Evening Post*.

The purpose here has been to show the imposition on the American public which is being practiced every day in the news columns of daily newspapers and other publications, but I have been able to convey to the reader only the barest kind of suggestion as to the depths to which this perception is practiced. Limitations of space prohibit further encroachment, or I would fain extend my list of examples indefinitely.

We hear much these days about the abuses of journalism. Much of the criticism is leveled at publishers who lend the use of their columns for boosting that is calculated to help their advertisers. But little attention is paid to that other evil namely, the use of the news columns for the purpose of destroying business rivals, political rivals and enemies generally of men who wield sufficient influence to employ the method.

This ramification of the subject appeals to me as of at least as much consequence to citizens as the one of inspired puffery. I believe the public is going to hear much more of this feature of newspaper abuse in the future than it has in the past. The community is waking up and is manifesting a desire to learn more about the heinous practice.

RAWHIDE AGAIN

To RETURN TO RAWHIDE. As a result of the scientific press agenting which the camp received, a frenzied stampede ensued. The rush was of such magnitude that it stands unparalleled in Western mining history. Not less than 60,000 people journeyed across the desolate, wind-swept reaches of Nevada's mountainous desert during the excitement. Not less than 12,000 of these remained on the ground for a period of several months.

Mining camp records were broken. The maximum population of Goldfield during the height of its boom was approximately 15,000, but it had taken more than three years and the discovery of the world's highest grade gold mine to attract this number of people. Cripple Creek for two years after its discovery was little more than a hamlet. Leadville during its first year was hardly heard of.

The scenes enacted in Rawhide when the boom was at its height beggar description. Real estate advanced in value in half a year in as great degree as Goldfield's did in three years. Corner lots on the Main Street sold as high as $17,000. Ground rent for plots 25x100 feet commanded $300 a month. During the day as well as at night the gaming tables of the pleasure palaces were banked with players, and the adventuresome were compelled literally to fight

Aided by spirited promoters like George Graham Rice, Rawhide in 1908 boasted a population estimated at 6,000 to 8,000 people. More than 40 saloons lined the main streets. The arrival of each stage brought considerable excitement

Announcement

THE RAWHIDE ROYAL-REGENT MINES COMPANY

—OFFERS—

50,000 Shares of Treasury .... Stock at ....

15 CENTS PER SHARE

The Proceeds of which will be used for development

THE Rawhide Royal-Regent Mines Company owns in fee simple, without debt or incumbrance, two full claims adjoining the estate of the Rawhide Rector Mining Company, the stock of which is listed at 50 cents.

It is a mining certainty that the continuation of three of the veins that are yielding leasers high grade shipping ore on the Rector estate, traverse the acreage of the Rawhide Royal-Regent Mines Co. The persistency of values along these ledges would make it appear certain they extend through the property of the Rawhide

Tex Rickard's Northern was Rawhide's most popular saloon

Northern Saloon
INC.
"TEX" RICKARD & Co., Proprietors

Schlitz Beer on Draught

Imported and Domestic Cigars

No. 2 NEVADA STREET, RAWHIDE, NEVADA

their way through the serried ranks of onlookers to take a hand in the play. The miners were flush. Many assay offices, accessories of high-graders, were turning out bullion from extraordinarily rich ore easily hypothecated by a certain element among the men working underground.

The opening of "Tex" Rickard's gambling resort in Rawhide was celebrated by an orgy that cut a new notch for functions of this kind in southern Nevada. The bar receipts aggregated over $2,000. The games were reported to have won for Mr. Rickard $25,000 on the first day. Champagne was the common beverage. Day was merged into night and night into day. Rouged courtesans of Stingaree Gulch provided the dash of femininity that was a prerequisite to the success of the grand *bal masque* that concluded the festivities. For the nonce social cast was utterly obliterated. Months afterward, when flames wiped out the commercial section of the town, the leading dance hall of the camp, situated a stone's throw from "Tex" Rickard's saloon, at the foot of Stingaree Gulch, was preëmpted without the slightest hesitation and promptly converted into headquarters for the distribution of relief supplies.

On the densely crowded streets fashionably tailored Easterners, digging-booted prospectors, grimy miners, hustling brokers, promoters, mine operators and mercantile men, with here and there a scattering of "tin horns," jostled one another and formed an ever shifting kaleidoscopic maelstrom of humanity.

In the environing hills could be heard the creak of the windlass, the clank of the chain, and the buzz and chug of the gasoline hoist, punctuated at frequent intervals by sharp detonations of exploding dynamite.

Outgoing ore laden freighters, hauled by ten-span mule teams, made almost impassable the roads connecting the camp with nearby points of ingress. Coming from the opposite direction, heavily laden wagons carrying lumber and supplies, and automobiles crowded to the guards with human freight, blocked the roadways.

Rawhide's publicity campaign from a press agent's standpoint was a howling success. From the standpoint of the promoter, however, results were mixed. Nat C. Goodwin & Co. were enabled to make more than a financial stand-off of their promotion of the Rawhide Coalition Mines Co., but they did not profit to the extent they might have, had the times been propitious.

I was not long in discovering that my first deductions, made at the inception of the Rawhide boom, namely, that the country was in no financial mood to consider favorably the claims to recognition of a new mining camp, were right, and that it would have been better had the birth of Rawhide been delayed for a period or until the country could catch its financial breath again. Crowds came to Rawhide, but few with money. Flattering as was the

Deep cuts in the earth and shaft headframes are all that remain of Rawhide, George Graham Rice's promotional paradise.

extent of the inrush, it was easy to see that if the publicity campaign had been suppressed for a while, the result in harvest would have been immeasurably greater. Had financial conditions been right, the effort to give the camp scientific publicity would undoubtedly have been crowded with results for the inside of a character that would have meant much larger sums of money in the bank.

Nat C. Goodwin & Co. recognized, too, that they had been working at a great disadvantage by attempting to finance a great mining enterprise at so great a distance from Eastern financial centers as Reno. We were hardly a match for the Eastern promoter who, because of the handy location of his offices, was enabled to keep in close personal contact with his following.

The usual happening in mining took place at Rawhide. The extraordinary rich surface deposits opened up into vast bodies of medium and low-grade ore at depth. Rawhide's one requirement appeared to be a railroad, and a milling plant of 500 or 600 tons a day capacity. It was decided that I should come East and attempt to finance the company for deep mine development, mill and railroad construction, and also to go through with the deal made with the vendors of the controlling interest. The time period for payments had been extended for Nat C. Goodwin & Co., and the option to purchase was now valued by the Goodwin company at a fortune.

In New York, over the signature of Nat C. Goodwin the firm for a while, under my direction, conducted a display advertisement newspaper campaign in favor of the issue, which was now listed on the New York Curb. Hayden, Stone & Co., bankers, of Boston and New York, who have since successfully financed the Ray Consolidated and Chino copper companies, undertook to send their engineer to Rawhide to make an examination of the property with a view to financing the company for railroad and milling equipment amounting to upward of a million dollars. Under the impetus of this news and the Nat C. Goodwin advertising campaign the market price of the shares shot up to $1.46, or a valuation in excess of four million dollars for the property.

A few weeks later a sharp market break occurred. Some one got the news before Nat C. Goodwin & Co. did that the million dollar financing proposition had been acted upon adversely by the engineer. The company had done no systematic underground development work. An enormous amount of work had been done, but it was accomplished under the leasing system. The leasers, who, because of lack of milling facilities, were unable to dispose of a profit of ore that assayed less than $40 per ton, had bent all of their efforts toward bringing to the surface high-grade shipping ore and had made no effort at all to block out and put into sight the known great tonnages of medium and low-grade. Engineers take nothing for granted and this one reported that the proposition of spending a million dollars should be turned down because a commensurate tonnage had not been blocked out and put in sight.

To this day the camp has struggled along without adequate milling facilities, but has been practically self-sustaining. From a physical standpoint the mines today are conceded to be of great promise. The company is honestly and efficiently managed. The president, from the day of incorporation to this hour, has been E. W. King, formerly president of the Montana Society of Mining Engineers, a director of a number of Montana banks, and recognized as one of the ablest gold mine managers of the West. M. Scheeline, president of the Scheeline Banking & Trust Co. of Reno, who ranks as the oldest and most conservative banker in the State of Nevada, has been treasurer from the outset.

The history of Rawhide is still in the making and its final chapter has not yet been written by any manner of means. Nor is it within the pale of possibility that such latent productive potentialities as have been established at Rawhide can long remain in great part dormant.

In Wall Street Nat C. Goodwin & Co.'s deal with the venders of the control of Rawhide Coalition was later financed to a successful finish. It was done by appealing to the speculative instinct of that class of investors who habitually gamble in mining shares. The effort to finance the mining company itself, to a point where it might take rank with the great dividend paying gold mines of the West, was not so successful.

"STRUCK IT RICH"

9. The Wall Street Game

A MAN WHO THINKS he knows what happened to me in Wall Street, and why it happened, suggests that the New York section of "My Adventures With Your Money" be prefaced with the following:

> This is the story of an energetic, self-confident, aggressive, optimistic, enthusiastic, nervy, fearless, imprudent, uncompromising, presumptuous *fool*.
>
> Maybe he was worthy of a following in that he would cast his own fortunes with those whom he asked to follow him, but withal he was a dangerous leader because he could not see the rents in his own armor and lacked caution, prudence and discretion. He could see a goal ahead and would lead the rush, but always failed to take into his reckoning one circumstance in his youth that left a blot on his escutcheon and placed in the hands of unfair opponents an envenomed weapon ready for use. He failed to see the necessity of making friends of his competitors and of placating his critics as he progressed. Indeed, he reckoned these elements not at all. He made many enemies and few allies. He never compromised. Naturally, he met with disastrous defeat for himself and the loyal ones who placed their faith in him.

I disagree. I was not a fool. I refused to be a knave, and I am not sorry. I have in mind a man of parts who as a stockholder has been doing the dirty work of unscrupulous multi-millionaire Wall Street mining promoters for years. Dishonest in his expressed opinions and a sycophant in his every action, the interest of the Wall Street man of power is always his as against that of the unprotected investor. I look upon this man as a vile person. I could not do as he does if my very existence depended upon such conduct. I would rather be out of business and broke for the rest of my life than be he. For me to serve the base purposes of high class crooks just because they have money and power, would be for me to barter away my soul and lose my peace of mind. I would not sell either for all the money in the world.

Honesty is the best policy. The type of man I have described can not thrive for long. He must evidently suffer total eclipse. The business of this world is founded and builded upon individual integrity. The business man who allows himself to be used to carry out the base purposes of men in high places forfeits the respect of those whom he serves, is forever afterwards mistrusted by them, and loses caste in the very set he tries to gain favor with.

I charge that powerful, dishonest interests on Wall Street found it necessary for selfish reasons that I be put out of business. I declare that they bided their time until newspaper clamor against so-called get-rich-quick promoters had been fostered, aroused and stimulated to a point where citizens became imbued with the idea that all promoters who use the advertising columns of newspapers are crooks. And I aver that when the government used upon me and my associates its rare power of seizure, search and confiscation, it was with no evidence that any government statute had been violated. In this and the concluding chapter of "My Adventures with Your Money" I state the facts which I believe prove these statements to the last syllable.

GOOD BIG FISH VS. BAD LITTLE FISH

Ask the casual newspaper reader to define offhand the compound adjective get-rich-quick and he will tell you it is applied solely to professional promoters who employ flamboyant advertising methods, promise great speculative profits, use other devices which are calculated to separate the public from its money, and are in every instance dishonest. That is the idea which powerful interests have inculcated in the public mind by subtle, insistent press agenting.

Time and again during the progress of "My Adventures with Your Money" I have endeavored to show that the really dangerous get-rich-quick forces are the men in high places who, by the artful and insidious use of the news columns of friendly publications and others which copy from them, divorce the public from millions upon millions. I said in my foreword the following:

> The more dangerous malefactors are the men in high places who take a good property, overcapitalize it, appraise its value at many times what it is worth, use artful methods to beguile the thinking public into believing the stock is worth par or

more, and foist it on investors at a figure which robs them of great sums of money. There are more than a million victims of this practice in the United States.

No man has right to assume that a promoter who sells stock by means of display advertising in the newspaper is *per se* a get-rich-quick operator. There are honest professional promoters of the display advertising variety and there are dishonest ones, just as there are honest promoters of the multi-millionaire kind and dishonest ones.

The on-the-level, trumpet-tongued mining promoter, who believes in newspaper advertising and successfully finances companies by appealing uproariously to the speculative investing public, performs an actual service and is entitled to a place among honorable men. Indeed, he is the hero of the prospector and poor mine owner of the West. He alone stands between these men and grasping monopoly.

Mine men, stockholders, and financiers the country over, understand this, although the Eastern newspaper reading public has been taught to believe that this type of promoter must be a get-rich-quick operator.

A broker in Wall Street who speculates in the securities of the New York Stock Exchange for his own account is considered unsafe. E. H. Gary, Chairman of the Executive Board of the Steel Trust, stated under oath at Washington in June that J. P. Morgan never speculates. Ask the average member of the New York Stock Exchange what chances the stock gambler has. If he is frank, he will shrug his shoulders and reply something like this: "If the game could be beaten, do you think I would be a broker? Wouldn't I be a player?"

The aggregate market value of seats on the New York Stock Exchange is nearly $100,000,000. It costs more than a hundred million dollars more every year to gather and transact through offices and branch offices the speculative business which forms the bulk of the transactions of the members. The kitty, or rake-off, is enormous. Who pays it? You hear of the stock broker going to Europe in his yacht every summer. How many of his trading customers travel that way?

Who pays the freight? Can a game be beaten where so many multi-millionaires are created among those who are on the inside and where so large a percentage of the speculator's money must come out every year to pay the enormous cost of maintaining a vast system of stock brokerage offices, stock exchanges, telegraph and telephone wires, newspapers, publicity bureaus, yachts, Fifth Avenue palaces, huge contributions to national and state political campaigns, etc.?

You hear a hue and a cry against bucketshops. There is no federal embargo against bucketshopping. Yet somehow or other the machinery of the government's Department of Justice is used to crush out this sort of gambling institution. Now, what is the difference in principle between gambling on margin on fluctuations of stocks in a bucketshop and doing the same through a New York Stock Exchange house?

This is the unimportant difference: The bucketshop keeper takes the other end of the play, pays you out of his pocket when the market goes your way and keeps your money when it goes against you. He never delivers any stocks.

The New York Stock Exchange member is expected to buy your stocks for you and *carry* them—some of them do and most of them don't, as is shown farther on—but in this case also no stocks are delivered to you.

The transaction is the same in principle as the one in the bucketshop, so far as the gambling feature is concerned. The only real difference is that when you gamble on market fluctuations through the bucketshops no contribution is made to the New York Stock Exchange kitty.

RIGHTEOUS WALL STREET AND THE SUCKER PUBLIC

THE NEW YORK Stock Exchange member will tell you that the evil of bucketshopping is that the bucketshopper is tempted when the public is long on stocks to depress the market by heavy short sales. On the other hand, the bucketshopper urges upon you that his business is gambling against fluctuations which he has no hand in making and that the financial powers of Wall Street resort to the same trick that he is occasionally accused of. The interests know at every hour in the day approximately how many shares of stocks have been borrowed for delivery against short sales or are being carried on margin for the long account. They know what the public's short interest or long interest is, and they, too, have it in their power to shake out the public at any moment they choose. Worse, it is common knowledge that this practice is continually resorted to. Stocks are put up and held up on bad news and marked down and held down on good news or no news at all. News is withheld and is manufactured to suit occasions. For years the market has been thimble-rigged to a frazzle. Margin trading suckers have been milked to a finish. George E. Crater, Jr., writes:

> Margin trading on the New York Stock Exchange is the most dangerous and destructive form of gambling known, because, being legal and therefore respectable, it allures hundreds of thousands of people who would never think of risking their money at faro, rouge-et-noir, roulette, or any of the other games

of chance. Statistics show that more people are ruined physically, morally and financially by stock gambling than by all the other forms of ordinary gambling combined. Monte Carlo is a Christian philanthropy compared with Wall Street. You have quite as good, if not a better chance to win a fortune at Monte Carlo than you have by putting up margins against Stock Exchange bulling and bearing, and if you ruin yourself at Monte Carlo the proprietor will at least refund enough of your money to pay your way home. The man who goes broke on margins finds no relief at his service on the Stock Exchange or among the brokers. There would not be so many millionaires in this country if there were not so many fools ready to throw their money away on margins.

A howl of condemnation is raised against horse racing. Newspapers, periodicals, politicians, enthusiasts, crusaders, and charlatans in every walk of life, are encouraged to make a big noise. Horse racing, like bucketshopping, is an avenue for speculation—gambling—and it keeps much money out of Wall Street. Fakirs, who are the tools of Wall Street, collect from Wall Street for their services and at the same time make moral or political capital of their zealousness in crusading against the various modes of gambling that create competition for Wall Street money.

The small fry mining promoter, who is not a member of the Stock Exchange, pays no toll to the big game, is beyond the discipline and control of the governing body of the New York Stock Exchange and is not a part of the machinery, sets up a competitive business which caters to the gambling instinct in the way of fluctuating mining stocks. The speculating public gets action, likes it, and invests money that might have been used in margin trading on the New York Stock Exchange or for investment in the constantly fluctuating low-priced industrials or higher-priced mining stocks that are sponsored by big interests with New York Stock Exchange affiliations.

Promptly the machinery of Wall Street is used to crush him. Column upon column is printed in the magazines and newspapers about get-rich-quicks. A conviction for crime is obtained of a real get-rich-quick offender— a little fellow who is guilty, but no more so than his licensed brother higher up, who is doing infinitely greater damage. The *one* that a coterie of high class Wall Street thimbleriggers are really after, because he thwarted them in their swindling operations by exposing them in his newspaper, but against whom they can not make a case, has a skeleton in his closet. They bring it forth, dangle it in the air, make the public think he, too, must be a scoundrel, and he is raided by a government agent during the uproar; and they get away with it. The righteous crusade against get-rich-quicks is press agented to the limit. The public falls for the dope. At last the government has acted to protect investors! Wouldn't it wilt you?

Were P. T. Barnum to be reincarnated and his humbugging mind by some miracle expanded a million times, it would still be impossible for him to conceive such a gigantic faking of the American public as it has been put to in the last few years.

And the public isn't on. Shrewd schemers on Wall Street keep pulling the wool over the eyes of the sucker public, and not only see no reason why they should discontinue the practice but find it very lucrative to continue doing it.

THE MARKETING OF MINING STOCK

AS A RULE, it takes much money to make a paying mine out of a promising prospect. Later on in the mine's progress, through the constructive period, other very large sums are generally required to pay for the blocking out of an ore reserve and to supply milling facilities for the reduction of the ores.

The peripatetic mining prospector of our Western mining empire—the dauntless finder of mines who laughs at hardship and ridicules the thought of danger, who makes companions of gila monsters and the desert rattler, whose only relief from the everlasting silence of the untrodden reaches of arid wastes is the sex call of the coyote—has the choice of just two markets for the sale of his find. He may either accept a comparatively small sum from the agent of a powerful mining syndicate for his prospect or he may receive a fair speculative price from the professional promoter.

The great mine financiers of this country rarely compete with one another for the purchase of any mining property. This is particularly true if one of the others happens to be operating in the district where the small mine owner's property lies.

As a rule, the original owner, whose entire fortune is perhaps tied up in the property, then finds himself in the position where he must either accept the first offer, however small, which is made to him by one of these dominant interests, or find that market closed to him.

His alternative, as mentioned, is a sale to the independent mine promoter of comparatively small means, who incorporates a company to own and develop the property and finances the operation from start to finish by selling stock in the enterprise to the general public.

The method of this class of professional promoter—the hope of the small mine owner—in marketing stock, usually involves the liberal use of the advertising columns of newspapers. He lacks pull or power sufficient to get his stock and mine talked of favorably in financial literature of the day to a degree that will excite public interest, and so he must construct his own publicity forces.

Advertising costs money and the public foots the toll. But if the promoter is honest, this item of cost is not in itself an argument in favor of stock offerings of the multi-millionaire mine capitalist who does not patronize the display advertising columns of the newspapers. Nor does it establish a case against the wares of the promoter who does. The promotion expenses incurred by the advertising promoter do not nearly approach in their totality the difference between cost price and the price at which the magnate promoter usually invites the public to participate in similar enterprises.

For example: a few years ago a certain man bought a certain mine for $1,000,000 on time payments. He has been making a market for the stock of that mine on the New York Curb at an average of above $8 per share, or more than $8,000,000 for the property. His firm, members of the New York Stock Exchange, have been advising people in their widely circulated market literature to buy the stock at this figure. And yet the property is without a reduction works, will need $2,500,000 to $3,000,000 in excess of money now in the company's treasury to erect one, which money must yet be raised somewhere and somehow, and the producing era of the company can not possibly begin for two years yet at the very earliest. I could cite many such instances.

When Nat C. Goodwin & Co. of Reno purchased the control of Rawhide Coalition, during the exciting Rawhide camp boom early in 1908, the valuation agreed upon for the property was $700,000. This was considerably more than the original owners could obtain for it at that time from any big interest. It, too, needed milling facilities.

As a matter of fact, but for the success of mine promoters of the Nat C. Goodwin & Co. and B. H. Scheftels & Co. class, the great Comstock Lode, which produced more than $600,000,000 in gold and silver bullion, would have likely remained undeveloped. The big public demand in the early 70's for Comstock mining shares of all descriptions was created by a series of flamboyant flotations and aggressive stock market campaigns. If the Con. Virginia mine had not opened up into a bonanza ore body at a depth of 1,400 feet, the frenzy of speculation in Comstock shares might have gone down in history as another South Sea bubble.

The brass-band promoter, be it understood, is therefore not without honor in the Far West. Deprive the mine prospector of the services of this style of enterprise projector, with his operating machinery, namely, facilities for appealing to the speculating-investing public, and you hit the small Western mine man a solar-plexus blow. Conversely, every obstacle placed in the way of the mine promoter of loud methods and moderate means is an added cause for rejoicing on the part of the Wall Street multi-millionaire mine capitalist.

When B. H. Scheftels & Co., with whom I was identified, were raided by the United States government in September 1910, a wail went up from the Western mine operator to his Representative in Congress. The best sentiment of the Far West, as I was able to gather it, favored the idea that the last hope of the small Western mine owner had been shattered. During the short period of B. H. Scheftels & Co.'s activity in New York it raised directly nearly $2,000,000 for Western mining properties and indirectly influenced in that direction at least $10,000,000 more.

The raid was a body blow to the small Western mine owner who needs capital to develop his properties and has no affiliations with capitalists. Since the raid I do not know of a mine owner of any of the great Far Western states who has successfully financed a mining proposition in the East except by delivering his property in its entirety into the hands of some big interest, which has taken it over for a sum insignificant by comparison with what the public may ultimately be expected to pay for it when the stock is finally marketed on the curbs and exchanges.

I BUCK THE WALL STREET GAME

AFTER I HAD CONDUCTED the big camp publicity campaign of Rawhide, which I had done with a view to centering the attention of the American investing public on the speculative possibilities of the stock of the Rawhide Coalition Mines Co., and in that way endeavored to finance the proposition—after I had failed by this method, in the teeth of the bankers' panic of 1907-8, to dispose of enough stock to finance the company for deep mine development, mill equipment and the payment to the original owners of the price for the control agreed upon, I came to New York, late in October 1908, bent on trying to succeed in the encompassment of my original purpose both by direct appeal to the public through display advertisements in the newspapers and by making a deal for part of the enterprise with the big fellows.

I found Rawhide Coalition stock listed on the Curb, and the market quiescent. Public interest in the East had been aroused to some degree, but the market was not absorbing stock. An effort to induce leading stock brokers to mention the issue favorably in their market letters failed. Those who were willing to give the stock some publicity exacted either a call on stock at a low price or an out-and-out reduction below the market quotation for such stock as they disposed of.

Such concessions were not to be thought of. It was the intention of Nat C. Goodwin & Co. to support a rising market for Rawhide Coalition. My Goldfield experience with mining stock brokers convinced me that few might be expected to protect the shareholders' interest in such an enterprise. Commission mining stock brokers of that period, who put their customers into a stock at, say, 30, were tempted to advise profit taking when the price advanced to, say 50, because by the operation they made another commission and often earned an additional, or third, commission by getting their customers out of the stock at a profit and into another one, levying a commission on each transaction.

Nat C. Goodwin & Co. decided to "try it on" direct with mining stock speculators by appealing to them through the advertising columns of the newspapers, asking them to purchase the stock on the New York Curb through their own brokers. Also, Hayden, Stone & Co., the Boston and New York banking firm, were induced to agree to raise $1,000,000 for the company for railroad and mill purposes, if their engineer would report favorably.

Provided with money with which to buy advertising space and furnished with stock certificates to supply the market, Nat C. Goodwin & Co. inaugurated an active campaign on the New York Curb.

What happened will be found instructive to the reader in several particulars; among them these:

(1) The free lance mining promoter does not always get the money when he succeeds in creating a buoyant market for his stock.

(2) Some stock brokers of seemingly high standing would just as soon skin a mining promoter of this order as they would an ordinary speculator. They play no favorites.

(3) Be a mine promoter ever so honest, without New York Stock Exchange affiliations his motives are bound to be misconstrued if he makes an error. The big fellows will sick on to him the newspapers or newspaper men whom they control or influence. Dust will be thrown into the eyes of the public so they'll buy the big fellow's fares, principally for sale on the New York Stock Exchange, and may forever be prejudiced against that of the little fellow's.

The campaign in Rawhide Coalition made good progress. It was early in November 1908. For six weeks I had been supporting the market for the stock on the New York Curb for Nat C. Goodwin & Co. of Reno. My office was an apartment in a Fifth Avenue hotel; our brokers were members of the New York Stock Exchange. For a month we had used, every day, display advertisements in the financial columns of New York City daily newspapers, signed by Nat C. Goodwin, to boom the stock. About 600,000 shares of the stock were in the hands of the public. The market, which was on the New York Curb, was real. Speculative buying had carried the price from 40 cents up to $1 a share. Mine reports were rosy. Wide distribution of the stock was taking place.

The public evinced deep interest. The Nat C. Goodwin advertisements set forth that $2 ought to look reasonable for the stock by Christmas day. There were reasons. Several very promising mines had been opened up. An engineer of high rank was examining the property. If his report should be favorable, a deal was practically assured that would involve the expenditure of $1,000,000 for deep mine development, a railroad, and adequate milling facilities. This, in turn, would mean early dividends for stockholders. Experienced, conservative mining men had expressed the opinion that the property bore the unmistakable earmarks of a big producer.

The stock became the feature of the Curb market. It easily occupied the center of the stage. Not less than 20 brokers could be counted in the crowd executing orders at almost any hour during the daily session. The fact that a New York Stock Exchange house was executing the supporting orders from the inside impressed the talent. Public buying through other New York Stock Exchange houses further convinced Curb veterans that the stock was "the goods." Up went the price under the impulse of public buying. Curb brokers themselves caught the infection. By December 7th the price soared to $1.40 per share. This was an advance of 500 per cent over the low for the stock of half a year prior.

THE DOUBLE-CROSSING OF RAWHIDE COALITION

AT THE CLOSE of the day's business on December 7th, our brokers, a single firm, members of the New York Stock Exchange, reported the purchase of 17,100 shares in the open market at an average price of about $1.39, and the sale of 1,800 shares at a little above this average. For the first time in the campaign there appeared to be selling pressure. We had quit long 15,300 shares. The sum of $21,000 in cash was required to pay for the long stock.

On December 8th, the day following, the same firm of brokers reported that they had purchased 17,800 shares at an average price of $1.37½, and the sale of 12,800 shares at an average price of $1.40—long on the day 5,000 shares.

On December 9th our purchases through this firm aggregated 16,800 shares at an average price of $1.40, while our sales totalled only 6,400 shares at a slight advance.

Nat C. Goodwin & Co. were now long on the three days' transactions 30,700 shares and had been called upon to throw $43,000 behind the market to hold it. This was a comparatively small load to carry and did not alarm us. We considered the stock worth the money. We were curious, however, to learn the reason for the selling.

Nat C. Goodwin & Co. had placed most of the outstanding stock direct from Reno with the investing public at from 25 cents to $1 per share, and early buyers were reaping a harvest. But this did not appear to be the explanation for all of the selling. Interest in the stock was now widespread. There was free public buying and for every actual profit taker there appeared to be a new purchaser. Apparently, somebody was selling the stock short.

Late that night a member of our brokerage firm which had been executing our supporting orders, called on me at my apartment. I inquired of him what protective orders he thought the stock would need the next morning to guard against professional attack. He replied:

"I think if you will give us a buying order for 5,000 shares at $1.35 there will be no difficulty."

My understanding was that he wanted to handle the market for me the next morning and that he would, of course, give me quick notice if further supporting orders were needed.

The order was given. It was a very ordinary precaution, for there is hardly a stock on the list that would not be raided by professionals if supporting orders were not known to be in the market. As Saturday is only a short two hours' session, I really fell in with the idea.

Retiring late that night, I left a call for 11 A.M. Next morning at about 10:45 I was awakened by my valet. He said Nat C. Goodwin wanted me on the long distance. Mr. Goodwin was in Cincinnati, where he was playing a week's engagement.

"Hello," said Mr. Goodwin. "Did they get you? Shall I wire the Knickerbocker Trust Co. to pay you $25,000 to support the market? Reported here they have you in a hole."

"What's up?" I inquired.

"Why, brokers here say the stock broke to 60 cents on the Curb soon after opening," he said. This was news to me.

"I do not need more money," I answered. "I have been asleep. Our brokers have been on the job. I will see what is doing and let you know in a little while. Don't worry." And I rang off.

I 'phoned our brokers and they reported that they had bought 5,000 shares of stock at $1.35 at the opening and had withdrawn support. "Too much stock was pressing for sale," they said.

"This is hell. You should not have permitted the market to break that way. Support the stock!" I said. "Buy 7,500 shares at the market!"

In a few moments this firm of brokers reported that they had rallied the market to $1.16. The recovery was only temporary, however. Another drive broke the stock to 60 cents.

Our brokers had bought 7,000 shares at from $1 to $1.16 and then stopped. The member of their firm who had been handling our orders throughout this campaign said the purchase of this fresh block of stock exhausted our cash balance on deposit with his firm. They had a number of drafts out for collection, attached to stocks sold to Western brokers, that had not yet been credited to us. There was also a big block of Coalition stock due us from them. This was the stock they had bought on our supporting orders. They refused, however, to consider either the drafts or the stocks as a credit.

We had cash on deposit and credit with a number of other brokers. I promptly telephoned several of them to buy large blocks of the stock at a limit of 95 cents. This was 35 points above the quotation that was given me. Not a single share was reported bought on these orders.

I jumped into a taxi and rode to the office of the brokers who had been handling our orders.

The situation was critical. I realized fully that a sharp break of this character in the market price of a stock that had been so widely exploited must prove shocking to investors. I feared that public confidence would be shattered completely.

"This is an outrage!" I protested. "Buy 5,000 shares at 95!" I tendered five $1,000 bills as payment in advance.

It was five minutes to twelve when I gave the order. At noon they reported that they had purchased 2,000 shares, for which I gave them the money. The market closed 95 bid for a wagon load.

On the face of things it appeared that the market had rallied from 60 to 95 on the purchase of 2,000 shares. This was another convincer that there must somewhere be much that was rotten about the play.

Investigation satisfied me that I had been double crossed.

The one firm of brokers, members of the New York Stock Exchange, who had been handling our orders, had acted as our clearing house, holding our stocks and our money. They had an advantage, which stock brokers understand well. Having executed most of our supporting orders, their agents on the Curb were also in a position accurately to judge the professional and lay speculation pulse. It was easy for somebody to put one over on us.

Shortly after noon I learned that Hayden, Stone & Co.'s engineer had turned down the proposition of advancing $1,000,000 for railroad and mill construction. A sufficient tonnage of ore had not been blocked out in the mine. Beyond a question this information was in the possession of brokers early in the day.

While I slept damage had been done to the market that was irreparable. By the time the price hit $1 on the way down trading had reached huge proportions. One clique of Curb brokers were reported to have been persistent sellers throughout. Their identity made it very plain that the double crossing process had been employed to a fare-you-well.

I accused our broker of not protecting our interests—the interests of stockholders. I raised a howl. He telegraphed another member of his firm who was away on a hunting trip, to come back to town. Next night both of these men, Nat C. Goodwin and myself met in my apartment behind closed doors. Their firm agreed to charge to their own account 3,000 of the 5,000 shares reported purchased for us at $1.35. Some other minor concessions were made.

On the day after the break New York newspapers reeked with sensational flub-dub about the causes of the smash in the price of the stock. In the preceding few months not less than a dozen other securities had busted wide open at various times on the New York Curb and New York Stock Exchange, but Stock Exchange houses were sponsors for these and the newspaper kept mum. Never on these occasions was there a hint in the newspapers that possibly someone had separated the public from its money.

Nat C. Goodwin and I were wrongfully accused of willfully smashing the market to shake the public out. The New York *Sun* printed an account of the break on the front page, top of last column. It began in a strain that indicated to confiding readers that chorus girls had lost their savings through the recommendations of Mr. Goodwin.

The *Sun* printed the list of officers of the Rawhide Coalition Mines Co., and emphasized the fact that I of Sullivan Trust Co. fame was second vice-president.

The *Sun* made no mention of the double cross. Nor did any of the other newspapers, with the exception of one.

The New York *Tribune* said:

A Stock Exchange house which has been putting out orders in the stock was charged with leading the attack on it yesterday, but members of the firm said that they had been acting merely as brokers for customers in the regular order of business.

Following the newspaper roasts, which helped further to destroy public confidence, two brokers on Logan & Bryan's continental wire system resorted to tactics of a kind to force lower prices. This wire has over one hundred out of town broker connections. A report was sent over the wire that Nat C. Goodwin & Co. had failed. Another followed it that the Rawhide Coalition Mines Co. was about to go into the hands of a receiver. The *Nevada Mining News* accused Nat Boas of San Francisco and J. C. Weir of New York of exchanging messages to this effect over the Logan & Bryan wire systems, so that all correspondents on the wire would have the false reports. Both Boas and Weir were believed to be short of the stock. Both were openly operating for a further decline. These and similar tactics resulted in a further easing off in price to 40 cents bid on December 24th, which was the low on the movement.

Two weeks after Christmas the stock rallied to 58 bid, 59 asked, and the market was firm again. On January 14 the price bulged to 70. At this point the stock again became the center of attack. By January 20 the price had eased back to 50.

Thus far the net result of Nat C. Goodwin & Co.'s various campaigns on Rawhide Coalition was the distribution of some 600,000 shares of stock. The issue had been well exploited. It had a big following and a broad market. Some excellent judges of mine values had become stockholders. The company, however, was still unfinanced for a long period of systematic mine development and mill construction.

We realized very clearly that some arrangement would have to be perfected to avoid a repetition of the trouble which the New York Stock Exchange brokerage firm had made us.

INSIDE MARKET SUPPORT

THE REMOVAL TO New York of B. H. Scheftels & Co., Chicago stock brokers, representatives there of Nat C. Goodwin & Co. of Reno, and a merging of brokerage and promotion interests of the two firms took place.

There was precedent for the move. There are a thousand other corporation interests in this country that are closely affiliated with Stock Exchange and other brokerage houses, through one or more of their directors or owners being partners in the business. As a matter of fact, it would be difficult to lay your finger on a single big interest of this kind that has not such a representation. These houses, of course, make it a rule to recommend the purchase of stocks in which their principals are interested. Affiliations of this kind are found essential to successful financing of enterprises. A number of New York Stock Exchange houses which are headed or controlled by men who are heavily interested in mining ventures that require financing are exponents of this method in the mining field.

Most of these have succeeded in promoting projects in which they or their associates are heavily interested, with the aid of the banking and brokerage facilities thus afforded. Principally by the use of the market literature and accompanying market manipulation, these houses have placed with their customers the securities of their firm members and associates. They have encompassed this by maintaining a brokerage, banking and promotion business without parading before the public, although never denying, the mixed nature of their business.

For the reader to comprehend the necessity for transacting the business this way, he should understand the underlying principle of financing an enterprise by the route of the listed stock market.

There are two ways of financing any enterprise with other people's money. One is by the primitive method of appealing directly to the public for subscriptions in huckster fashion, taking the money and then refraining from listing the stock or establishing an open market for it. You can't finance an enterprise of consequence these days by any such procedure. It is practically impossible to borrow from banks or from loan brokers on any security that has no fixed market value. A market must be established, for without a market on which to sell, intelligent investors won't buy.

The method, therefore, in common use, and the only one which has been found effective by financiers, is to create a demand for the security, encourage speculation, establish an active market, and dispose of stock on the market as necessity demands whenever financing is required. This implies and necessitates that the inside interests must support the security in the open market. Therefore, it becomes necessary for the successful marketing of the stock by the promoters, once a demand is created and public buying is under way, that stockholders shall be kept in full touch with the latest transpirations on the property and in the market—be furnished with news concerning their interests so that they may judge the value of their stockholdings. This process is particularly essential during the financing period of the company and the security digesting period of the public.

In fine, the ultimate purpose in this regard of all the promotion machinery of Wall Street—the machinery that has been putting out billions of dollars' worth of securities to investors—is to place stock where it will stay put, that is, not come out on the open market again to embarrass the interests that are behind the enterprise, and who for a long period are compelled to support the market.

On the question of the ethics of market support by the inside, a whole tome could be written. I will not attempt to discuss the subject at length here. Suffice it to say that in my opinion inside support of a listed security is not base when it is done with a view to creating a broad market, to stimulate public interest, and to increase the price to a point within the bounds of intrinsic plus reasonable speculative worth. Support of the market to the point of stimulation is moral obliquity, however, when dishonestly performed for the sole benefit of the inside and to the hurt of the stockholder. This sort of market support is only a shade less reprehensible than manipulation that has for its purpose the reduction of the market price of a security to beneath its real value, which, in my opinion, is nearly always infamous.

I might place myself on record right here to the effect that only once did I ever bear a stock from the inside, and on that occasion it was a temporary affair, caused by a desire to secure at a reduced price a big block of stock that was pressing for sale from a quarter that I was under no obligation to. Even in that instance I gave the investor much of the benefit my associates secured by letting him have stock at the same figure at which the inside secured it. Nor have I ever tried to push the price of a stock to a higher level than that which I considered warranted by the reasonable speculative and demonstrated intrinsic value behind the security.

B. H. SCHEFTELS & CO.
MINING STOCK BROKERS
Rector Bldg., Chicago, Ills.

We Buy and Sell
Rawhide Coalition
Rawhide Queen
Rawhide Consolidated

Listed Nevada Stocks Our Specialty.
Orders Given Instantaneous Execution.
Direct private wires in our office to all exchanges.
Daily Market Letter.

10. B. H. Scheftels & Company

B. H. SCHEFTELS & CO., INCORPORATED, mining stock brokers, successors to B. H. Scheftels & Co., for many years stock brokers in Chicago, opened its doors on Broad Street, New York, on January 18, 1909. For a long period B. H. Scheftels & Co. of Chicago had been advertised as the Eastern representatives of the corporation of Nat C. Goodwin & Co. of Reno, of which Mr. Goodwin had been president. It was now announced that Nat C. Goodwin had become vice-president of the new corporation of B. H. Scheftels & Co. Because Mr. Goodwin was by profession an actor and not a stock broker and because of the personal abuse he suffered in unfair newspaper criticism which followed the break in the market price of Rawhide Coalition a month before, he was quite willing to serve as vice-president instead of president. Besides, he could not spare the time from his profession to attend closely to the business.

The new corporation of B. H. Scheftels & Co. made its bow to the public by at once featuring in its market literature advice to purchase stock of the Rawhide Coalition Mines Co. I became publicity manager for the Scheftels corporation, manager of its promotion enterprises, and was placed in charge of the protection of the corporation's interests in all markets where its stocks were traded in.

Soon I was conducting a fresh campaign with investors that became so hot, so exciting and so big that for nineteen months I labored on an average sixteen hours a day, including Sundays, without being able to complete in a single day a day's accumulated business. The business grew until B. H. Scheftels & Co. were actually spending more than $1,000,000 annually for office and publicity expenses. In the nineteen months of its existence it bought, sold and delivered approximately 15,000,000 shares of mining stock. The Scheftels corporation broke every record in this regard that was ever made by a mining stock brokerage and promotion house in the history of Wall Street. Throughout its career it was viciously attacked from many directions, but it held its own. Through its hold on the mining stock speculating public, who were getting fairer treatment than ever before, it survived the concerted onslaughts of a number of important interests which it had competed with and antagonized, until one day in September 1910, on a warrant sworn to singly by one George Scarborough, since permitted to resign, clothed with the office and power of a special agent of the Department of Justice, its offices were raided, its books and papers seized, its property confiscated, and its officers and employees arrested.

The annual expense of B. H. Scheftels & Co. was $1,000,000 or more. Follows a tabulated statement of the expense item. The figures are approximated. The books of the corporation, which are now in the possession of the Department of Justice of the United States, will probably show that the annual expense was larger. The books not being readily available, an attempt is made here to be ultra conservative in setting down figures:

ANNUAL EXPENSE OF B. H. SCHEFTELS & COMPANY

Establishment of main office and six branch offices, furniture, fixtures, etc.	$ 40,000
Office rentals	35,000
Private wire system connecting branch offices in six cities with New York	25,000
Telephones	5,000
Telegraph tolls	100,000
Salaries (all offices)	200,000
Daily and Weekly Market Letter, printing and postage	100,000
General office expense, etc.	100,000
Miscellaneous postage	25,000
Miscellaneous printing and stationery	25,000
Advertising, publicity, etc.	200,000
Expert accountants	15,000
Commissions and salaries to Curb brokers	50,000
Mining examinations, engineers' fees, legal fees, etc.	50,000
Interest charges	30,000
Total	$ 1,000,000

Before the Scheftels corporation was in business a month it became plain that it was filling a long-felt want. In almost every branch it was performing some function in a manner more satisfactory to mining stock speculators and investors than were its competitors.

Its Market Letter news service, usually 16 pages, was the prime article. It soon gained a circulation of 34,000 among the highest class and best informed stockholders of mining companies in the country. It was also regularly sent to more than 2,500 stock brokers, including members of the New York Stock Exchange, New York Cotton Exchange, Boston Stock Exchange, New York Produce Exchange, etc.

Before the Scheftels corporation was five months old, the work of its Market Letter was supplemented by the *Mining Financial News*, a weekly newspaper which had been published for a long period at Reno as the *Nevada Mining News*, latterly as the *Mining Financial News*, and which removed to New York when the Scheftels company found the mining stock public was hungry for real live news and the truth regarding the mining propositions of other states as well as those of Nevada. The *Mining Financial News* and the Scheftels Market Letter, which were published three days apart, were supplied with news from practically the same sources. The newspaper was mailed to all readers of the Market Letter.

The ablest and most reliable mining correspondents obtainable for money in Tonopah, Goldfield, Ely, Rawhide, Cobalt, Butte, Globe and other mining camps, and the most experienced market news gatherers in the mining stock market centers of Salt Lake, San Francisco, Boston, Philadelphia, Toronto and New York, were placed on the payroll. Brokers in these and other cities, including Duluth, Seattle and Butte, supplied more news.

Wherever there was mining or market activity, representation of the very highest character was sought. News was always wired, no matter what the cost, whenever it was important to traders in mining shares. Expense was never spared when the information was considered of value to the speculator or investor. In the New York offices of the Scheftels corporation and the *Mining Financial News*, which adjoined each other, a staff of newspaper men with a mining financial experience of years was gathered. Little that transpired in the mines or the markets ever got away from them. Days before the mining newspapers of the West reached the East the Scheftels Market Letter or the *Mining Financial News* communicated the news regarding mine developments. They also contained a daily and weekly stock market diagnosis and prognosis. These were based on the news, as gathered by trained forces and aided from time to time by secret information which filtered into the offices. This service soon obtained an accuracy theretofore unknown on the Street.

There is probably not one stock broker in five hundred that would know a mine underground if he saw one. On the payrolls of B. H. Scheftels & Co. and the *Mining Financial News* there were thirty men who had been literally brought up in the mines and who, when they put their pen to paper, knew what they were writing about. The Scheftels company and the newspaper furnished mine and market information of quality to investors who had before been inundated with misinformation, guesses and twaddle. It sought to guide mining stock speculators right.

It was really a delicate job to handle the *Mining Financial News* in a manner which would not lead stupid people to believe that it was an entirely independent paper. It was desirable that its independence be maintained to a degree, so that the full value of the *Mining Financial News*, as a property, might grow. The intention was some day, when the *Mining Financial News* found itself on a paying basis, to sever the Scheftels alliance.

The *Mining Financial News* had always been an entity. It had up to then been assisted financially at periods by mining promotion concerns with which I had been identified and was always a quasi house organ for this reason. But it invariably preserved a certain independence in its news columns and at least such partial independence of ownership as enabled it to stand on its own bottom.

MORE TRUTH ON THE "MINING FINANCIAL NEWS"

WHEN THE *Mining Financial News* removed to New York, Mr. Scheftels used much persuasion to get the owner to transfer title to the Scheftels company. Admittedly, if the Scheftels company could boast ownership of the newspaper at the head of its editorial page, it would be a great feather in the Scheftels cap and might lead investors to think that an organization which could own and publish a first class, metropolitan newspaper of the *Mining Financial News* variety must for that reason alone be worthy of financial credit.

Thompson, Towle & Co., members of the New York Stock Exchange, print a small sized pattern of such a newspaper, called the *News Letter*. Hayden, Stone & Co., and Paine, Webber & Co., of Boston and New York, are said to have much influence with the Boston *News Bureau*, a newspaper which features news of mines and mining share markets. The Boston *News Bureau* at times has printed no display advertisements and at other times has. It is considered by Boston mining stock brokers who handle the Michigan and Arizona copper securities as a necessary complement to their market literature.

Walker's Copper Letter and the *Boston Commercial* are other examples. *Walker's Copper Letter*, which carries no advertising, for years has said the very nicest things about copper securities promoted and fathered by important Boston and New York interests. Needless to state, what *Walker's Copper Letter*, the *Boston Commercial* and the *Boston News Bureau* say about the mining propositions of their friends is as a rule based on fact. The point is that promoters find it necessary that news happenings regarding the markets, the securities and the mines in which they are interested be given broad publicity.

It was the idea of the owners of the *Mining Financial News*, of which B. H. Scheftels, president and 25 per cent owner of the capital stock of B. H. Scheftels & Co., was not one, that anybody who would supply the sinews while the paper was getting on its feet and was establishing itself, was entitled to all the publicity which the paper could consistently and honestly give it. With this understanding the Scheftels company assumed to take all of the income of the *Mining Financial News* and pay all of the running expenses until such period as the newspaper might become self-sustaining.

In doing so it performed a stupendous service to the entire mining industry in that the space devoted to the Scheftels enterprises therein did not average more than one-eighth of the whole, and it spent dollars to supply the news of all stocks where other mining financial publications in its field spent pennies.

To make sure that the public understood the *Mining Financial News* was the quasi house organ of the Scheftels company many precautions were taken. No application was made for admission to the mails as second class matter, and the paper was mailed under one and two cent postage. The name of Harry Hedrick was lifted to the top of the page as vice-president of the corporation owning the *Mining Financial News*, Mr. Hedrick being openly employed by the Scheftels company as head of its correspondence department. My own name was later placed at the head of the editorial page as editor, the Scheftels company making no bones about my position as absolute head of its publicity department, its promotion enterprises, and of all markets for the Scheftels promotion stocks. The connection had before been established even closer than this. I had formerly been advertised as vice-president of Nat C. Goodwin & Co. of Reno and vice-president of the Rawhide Coalition Mines Co.; and the Scheftels company had advertised that Nat C. Goodwin was its own vice-president.

Further, the Scheftels company announced in its market literature that it had selfish interests in protecting the market for the stock because of the Nat

C. Goodwin affiliation. Occasionally market articles under the signature of B. H. Scheftels were published on the front page of the *Mining Financial News*. Whenever anybody made a request for the Scheftels Market Letter a copy of the *Mining Financial News* was quite regularly mailed to him without cost. Articles under the signature of other officers and employees, formerly of Nat C. Goodwin & Co. of Reno and later of B. H. Scheftels & Co. of New York, were very frequently printed in the *Mining Financial News*.

Probably the most important reason why the Scheftels company made this sort of arrangement with the *Mining Financial News* was that it could do so with only a very small additional outlay. The Scheftels company found it necessary to employ correspondents in all mining and market centers, and the same correspondents could work for both enterprises. Another economic argument was that an enormous saving could be made in telegraph tolls, all dispatches addressed to the newspaper being sent at press rates. These dispatches were always available to the Scheftels corporation and its clientele.

It was the idea of the Scheftels organization that the mining stock investing public sorely needed right direction and that any brokerage house which led it right would soon be unable to transact all the business that would be offered to it.

And that is just what happened. Before the Scheftels company was six months old the fifteen men in its accounting department were compelled to work day and night—time and again throughout the night until 6 A.M.—to catch up with their work.

If the Scheftels news service was as nearly perfect as money and brains could make it, its facilities for the execution of orders on the New York Curb, the Boston Curb, San Francisco Stock Exchange, Salt Lake Stock Exchange, Toronto Stock Exchange, and other mining markets were unsurpassed. Its New York and Boston offices were connected with branch offices in Philadelphia, Chicago, Detroit, Milwaukee and Providence by exclusive private wires, and the service to out of town offices was almost instantaneous.

The New York offices were located right in front of the Curb market on Broad Street on the ground floor of the big Wall Street Journal building, 50 feet by 200 feet deep—occupying about 10,000 square feet of floor space. The Boston office, occupying two floors, was located within 100 feet of the Curb market in that city. The public wires of the telegraph companies gave quick service between San Francisco, Salt Lake and Toronto, where business was transacted through members of the mining stock exchanges of those cities. The private wires of the Scheftels company were constantly flooded with rapid quotations and market, mine and company news during every trading hour. In New York the Curb brokers in the Scheftels employ, some on salary and some on commission, rarely numbered less than ten and at one period exceeded twenty.

The correspondence department was presided over for a long period by two of the best posted mining market men that could be employed for money. From this department were usually graduated the managers of out of town offices. In the cashier's cage six men were engaged at an average salary of above $100 a week, registering stocks, receiving stocks, paying money and drawing checks. The payroll of the mailing department, which was operated in conjunction with the *Mining Financial News,* was comparatively small. Money saving machinery for the handling of the large output of market letters and newspapers gave excellent and economic service. About ten stenographers were regularly employed in the correspondence department. Occasionally, when a special effort was being made to interest the public in some security in which the corporation was particuarly concerned, a force of forty additional typists was pressed into service for short periods.

THE SCHEFTELS PRINCIPLES

WHEN THE CORPORATION of B. H. Scheftels & Co. opened its doors in New York it had no affiliations with any other Wall Street interests. It had no axes to grind except its own. It was practically a free-lance. It cracked up its own wares, careful always to keep within the facts, and never minced words about the quality of the goods of its contemporaries. The principle of both the Scheftels corporation and the *Mining Financial News* was to be always *right* in their market forecasts. The general order to mine and market newsgatherers and market prognosticators was to give the facts.

The law laid down was this: If the news is bad and is likely to injure the interests of our best friends, tell it in the interests of the investor. If it is good and the backers of the stock affected happen to be our worst enemies, tell it. No matter on what side of the market you think B. H. Scheftels & Co. is committed in any of its own speculations, give the customer all the news. Put the cause of the mining stock trader in front of you as the one to further always. Never exaggerate. Eventually, this policy must redound to our credit and profit.

Eventually, this policy resulted in our ruin. Our truth-telling policy was directly responsible for the loss of millions to competing promoters, and they banded together to destroy us.

The publicity, promotion and brokerage activities of the corporation were of such magnitude, and withal so simple, that they at once challenged the attention of the Street. Before the Scheftels corporation was half a year old, veterans of the financial game began to opine that some big interest was behind the concern. Its dashing market methods, its mighty publicity measures and its unbridled assurance attracted much notice. From every quarter expert views reached the Scheftels company that its manner of doing things was convincing on the point that it knew the business. But the general opinion of the talent seemed to be that the new corporation was spending too much money and that it could not win out unless a big boom in mining shares ensued.

The market tactics adopted by the Scheftels company in its promotion enterprises were as old as the hills. On the New York Stock Exchange they had been employed in a thousand instances before. The method will probably survive all time. The corporation sought to distribute the stocks of which it became sponsor in turn—first Rawhide Coalition, then Ely Central, later Bovard Consolidated and finally Jumbo Extension—by the approved Wall Street system of establishing public interest and inquiry and causing an active market. The aim was to establish higher prices for the securities, always within the bounds of intrinsic and reasonable speculative value. All efforts were directed this way.

Plans like this are, however, sometimes thwarted. Markets get sick. More stock presses for sale than the inside has money to pay for. Stocks break in price. Then the promoter can't make any money and might lose a lot of it. Since money making is his primary object, and stock distribution secondary, he has got to do some close figuring when markets are subject to the price breaking habit. That's where B. H. Scheftels & Co., through its brokerage business, found, after a short period, that it held within its grasp the power to insure itself against declining markets.

Without promotion stocks on hand—obtained by wholesale at lower figures than values warranted—in which it could profit to the extent of hundreds of thousands of dollars on a rising market, the million dollar annual expense of the Scheftels company would not have been justified. Once the market sought lower levels and no profit could be made on the promotions, it meant a discontinuance of the business on the large scale.

The corporation's insurance was the open market in stocks on the general list and its brokerage business.

From time to time it openly shorted tens of thousands of shares of stocks in which it had no promoter's interest whatever, by going out in the open market and selling them to all bidders against future delivery, by borrowing them from brokers and selling them for immediate delivery, and by short sales generally.

Speculators play the market and so did the Scheftels company, but never against its own stocks. Speculators, however, buy mining shares outright or on margin because they want to gamble. The Scheftels company played the market for just the opposite reason. It didn't want to carry its eggs in one basket and wanted insurance against market declines to cover promotion losses that must ensue if a general market slump occurred.

And the Scheftels company did not inaugurate any fake bookkeeping system or otherwise hide behind any bushes in doing this.

Morever, the corporation didn't take advantage of anybody. The cards were not marked. The deck was not stacked. There was no dealing from the bottom. Market opinion for which the corporation was directly or indirectly responsible was genuine to the last utterance. No news was suppressed on any stock. The corporation divulged to its customers and to the general public every piece of important outside or inside information regarding any stock on the general list that was in its possession. At the very moment when it was going short of stocks in greatest volume its market prognostications were winning for it a reputation for accuracy never before recorded.

If the stocks which the corporation went short of—stocks on the general list and amounting to probably 15 per cent of the volume of its entire business, the remainder of the transactions being all in house stocks (these house stocks it could not be short of because of its promoter's options on hundreds of thousands of shares)—if the stocks on the general list thus shorted went up in price and the corporation was compelled to go into the market later and cover at a great loss, it was always in the corporation's heart to sing a pæan of thanksgiving, for it could well afford to pay the losses sustained by it in the general list out of the greater profits which would be made in the house stocks, which must, forsooth, share in the general upswing.

Collateral securities put up by customers as margin for the purchase of other stocks were credited to the customers' accounts and mixed with the company's own securities. In every case proper endorsement of certificates, put up for collateral margin, was required. Every certificate of stock bears on the reverse side a power of attorney, in blank. The signature thereto of the person to whom the certificate was issued makes it negotiable by the broker. It was the rule of the house always to inform those who brought collateral to the offices for margin that the stocks would be used and that they would not receive the identical certificates back again. In a number of cases objection was made. Acceptance of the stock as collateral margin was then promptly refused. If there were any scattering exceptions to this rule, it was contrary to instructions and due to neglect or ignorance. Whenever a customer closed his account and demanded the return of his collateral, stocks of the same description and denomination were recalled and delivery made.

The same rule applied to stocks pledged with the corporation for loans, it being specifically set forth in the promissory note which the borrower signed that the privilege of using the stock was granted to the lender.

This practice is so common and the rule so generally understood by mining stock traders that objection was rarely made by customers.

To test the general custom, a friend at my suggestion not long ago sent certificates of stock to 17 stockbrokers now doing business on Wall Street. Three of these were members of the New York Stock Exchange and 14 were members of the New York Curb, Boston Curb, or of a mining exchange. A letter substantially as follows was sent to each of the 17:

> Enclosed please find ... shares of ... stock to be used as collateral margin for the purchase of an additional block of ... shares. Please buy at the market and report promptly.

The 17 orders were executed by the 17 individual houses. A month later when the stock ordered purchased had advanced in the market, the following letter was sent to each of the 17:

> Please sell the ... shares of ... stock which you purchased for me a month ago at the market and return to me the certificate of stock I sent you as collateral with check for my profits.

It took nearly two months for all of the 17 to make delivery. When they did, not one of them returned the same certificate that had been put up as collateral. Don't be shocked, dear reader, at this disclosure. It is the *custom*.

And don't, please, think mining stock brokers are alone given to the general practice. If you order the purchase of a block of stock on cash margin from any New York Stock Exchange house or send a certificate of stock as collateral in lieu of cash to one of them for the purchase of more stock, you will receive a confirmation slip of the trade which will generally read something like this: "We reserve the right to mix this stock in our general loans, etc." That is, the right is reserved, and actually exercised, of immediately transferring ownership of the certificates to the broker.

Unless a certificate stands in a customer's name and is unendorsed by him, he has no control over it. According to law, a broker has a right to hypothecate or loan securities or commodities pledged with him, for the purpose of raising the moneys necessary to make up the purchase price, and such stocks have no earmarks. In other words, the customer is not entitled to specific shares of stocks, so that stocks bought with one customer's money may be delivered to another customer.

As for the Scheftels company laying itself open to the charge of bucketshopping in shorting stocks, such a possibility was never dreamed of. The penal law of the state of New York, sections 390 to 394, inclusive, is the only criminal statute covering market operations commonly known as bucketing and bucketshops. In each section and subdivision it is provided that where both parties intend that there shall be no actual purchase or sale, but that settlement shall be made on quotations, a crime has been committed, the language of the statute being, "wherein *both parties* thereto intend, etc.," or "where *both parties* do not intend, etc." The Scheftels company was never a party to any such arrangement. And it always made it a practice to make delivery of stocks ordered purchased within a reasonable period after the customer had paid the amount due in full.

Now, neither myself nor the Scheftels corporation is responsible for brokerage conditions as they exist, nor for the laws as written. Custom and practice are responsible. The purpose here is to communicate the exact nature of the business methods of the Street as I found them and to lay particular stress on those that are open to criticism.

THE SCHEFTELS COMPANY AGAINST MARGIN TRADING

THE SCHEFTELS COMPANY did not encourage margin trading by its customers. In fact, it railed against the practice. Time and again the *Mining Financial News*, editorially, denounced the business of margin trading. The Weekly Market Letter of the corporation sounded the same note. On several occasions, in large display advertisements published in the newspapers, the Scheftels company decried the practice and urged the public to discontinue trading of this character.

There were selfish reasons for this. In the marketing of its promotions the Scheftels company found that not more than 20 per cent of the public's orders for these stocks given to other brokers were being executed, or, if executed, that the stocks were at once sold back on the market, the brokers or their allies standing on the trade.

Had the Scheftels company been able to destroy the practice by its campaign of publicity, it would undoubtedly have been able, during the nineteen months of its existence, successfully to promote three or four times as many mining companies as it did, and its profits would have been fourfold.

It, however, appealed to the public in vain. Loud, frequent calls to margin traders to pay up their debit balances and demand delivery of their certificates, which would compel every broker to go out in the market and buy the stocks he was short to customers, failed miserably.

The lesson of this experience was that the speculating public did not give a rap whether their brokers were short of stocks to them or not. All they wanted, apparently, was to be assured that when they were ready to close their accounts, their stocks, their profits or their credit balances would be forthcoming.

What is the evil of short selling of the kind described herein? The only evil that I could ever discover was that the market is denied the support which the actual carrying of the stock is calculated to afford. This hardship weighs heaviest on the promoter. There appears to be no cure. Even if a broker does buy the stock and does not himself sell it out again, there is no law that denies him the right to borrow on it or loan it to somebody else. And it is to the interest of the broker, because he gets the use of the money, to loan the stock always. Stocks are rarely borrowed by anybody except to make deliveries on short sales.

What about the broker who doesn't execute his order at all but stands on the trade from the beginning and sells the stock to his own customer, delaying actual purchase until delivery is demanded? This practice is even less damaging to the customer than the one of actually executing the buying order for the customer at the time the order is given and then selling the stock right back on the market again for the account of the broker or his pal—the usual practice when the object of going short is sought. When a broker buys stocks in the market he must bid for them, and actual purchase generally means a higher cost price to the customer than that at standing quotations.

The rule of the Street is to charge the customer interest on all debit balances. When a broker lends to a short seller a stock which he is carrying for his customer, he is paid the full market value, as security for its return. In that case the broker ceases to incur interest charges for the customer, and is actually able, in addition, to lend out at interest the cash marginal deposit put up by the customer.

Maybe you think, dear reader, that a broker who charges his customer interest at the rate of six per cent per annum on money which he has ceased to advance is crooked. Very well. If that be so, then all members of the New York Stock Exchange must be labelled crooks. Here is how it works, even among the highest class and most conservative members of that great securites emporium:

John Jones orders the purchase by his broker of 1,000 shares of Steel on margin. He pays down 10 per cent of the purchase price. Mr. Jones receives a statement at the end of the month charging him with interest at the rate of six per cent per annum, or more if the call money market is higher, on the 90 per cent of the purchase price advanced by the house.

On the same day that the order of John Jones is received, William Smith orders the same house to sell short 1,000 shares of Steel at the market. This order is also promptly filled. Thereupon the broker uses the 1,000 shares of Steel, which he bought for the account of John Jones to make delivery through the Clearing House for the account of William Smith. Sometimes a fictitious William Smith is created, known as "Account No. 1," "A. & S. Account," "E. Account," etc. This is usually done when a broker wants to hide from his bookkeepers that he or an associate is taking the other end of the customer's trade.

The broker is out no money, yet he charges Mr. Jones the regular rate of interest on his debit balance. As a matter of fact, too, the stock bought for Mr. Jones is never even delivered to his broker. The Clearing House, because of the short sale, steps in and delivers it to the broker to whom it is due "on balance." Custom and practice cover a multitude of remarkable transactions —don't they?

You have the framework of the Scheftels structure and of its Wall Street environment outlined in this chapter. Some of the narrative is undoubtedly dry as dust, but its recital has appeared to be necessary to enable the lay reader properly to interpret the chronology of stirring events which forms the concluding installment.

In the foregoing I have endeavored to lay bare many practices that are common to Wall Street. Wherever I have laid them at the door of B. H. Scheftels & Co., I have given that corporation much the worst of it, because in the recital I have omitted to mention a multitude of happenings that were creditable to an extreme to the Scheftels company. Most of these had to do with the experiences of the Scheftels company as publicity agents and promoters. Its wide open publicity and promotion policy called forth the ire of influential Wall Street pirates and caused the pressure at Washington which resulted in the federal raid of the Scheftels offices.

I have reserved this dramatic series of events for my last chapter.

CHANGE OF NAME BY RICE SCHEFTELS TAKEN AS A CLOAK FOR FRAUD

Court Rules That a Wolf Under Any Other Name Has Fangs As Sharp As Ever and That Goodwin, Rice, Scheftels, et al. Were the Same Concern.

Another week of the Rice-Scheftels trial and investigation of the methods adopted by the defendants brought out sworn statements of startling knavery. Stocks were bucketed, orders seldom filled when they could be avoided and mail order victims were hectored into letting their claims go the way the alleged brokers desired. The hearing is still on before Judge Ray of the United States district court, for the District of New York.

Judge Ray has admitted temporarily evidence relating to operations of Nat. C. Goodwin & Co., the name under which Rice was operating when he secured control of Rawhide Coalition. The contention of the government was that B. H. Scheftels & Co., was simply a continuation of Nat. C. Goodwin & Co., under a change of name.

The jury was excused while both sides argued the question of legality of evidence concerning the operations of Rice while operating as Goodwin & Co., before the organization of Scheftels & Co. The defendants contend that the indictments allege offences committed under the Scheftels regime, that they were not responsible for the sins of Goodwin & Co., even though Rice was a party to them. In the course of the argument the government contended that Goodwin and Rice, as already shown, were interested in the business of Scheftels & Co. up to the time of the raid, and that the contracts and wicked ways of Goodwin & Co. were transferred to the B. H. Scheftels Co. under the new name. The district attorney said that the scheme to defraud the public which was operated by the Scheftels concern, was conceived by the Goodwin company and that the change of name was only a cloak to conceal a well defined purpose to continue the swindle.

"The government purposes to show," said the district attorney, "that an agreement existed between the Scheeline Banking and Trust company, of Reno, Nev., and the defendants, to control the price of Rawhide Coalition stock. In other words that B. H. Scheftels & Co., controlled the stock, that they represented to customers that they had no interest in Rawhide Coalition but bought the stock in the open market, and that in addition to charging their own price they also charged commissions for themselves had to Banking & Trust Co. of Reno, was placed upon the stand and identified the agreement made with the Nat the agreement made with the Nat Goodwin company for control of Rawhide Coalition stock. The agreement shows 750,000 shares placed in the treasury of the latter company and 1,800,000 with the Scheeline Banking & Trust Co. The Goodwin company had an option on all the shares of the Rawhide Coali-

(Continued on Page 5.)

HIGH GRADE OUTPUT BY VICTOR LEASERS

The Hardwick and Lowden lease on the Victor claim of the C. O. D. is showing some very fine shipping ore. A panning made recently yielded nearly a half teaspoonful of gold, not the light, flaky particles, but good shot and grain, with sharp edges, showing that it has weight and substance. A catch sample from there yielded an assay of over $2000.

The lucky leasers are preparing a shipment, to go out soon. The ore is being extracted from the 300 and a winze from that level. A maintenance of this grade is apt to cause a sensation, more especially as last week it was considered that the shipment would run somewhere between $300 and $500 per ton. The working force is small, but it is composed of careful, conservative miners.

LEASE ON BLACK BUTTE PRODUCING HIGH GRADE

CAMPBELL NOW ACCUMULATING A SHIPMENT TAKEN FROM THE 100 LEVEL.

The Campbell lease on the Black Butte ground at Diamondfield is just now enjoying a high grade streak of shipping ore, and it is not such a little streak, either, for it is a good two feet wide. This is on the 100 level, reaching out from the Dortch shaft,

11. A Fight to the Death

In professional quarters the Scheftels corporation was regarded as an interloper from the day it set foot in the financial district.

Its first offense was to reduce its commission rates. This move set the whole Curb against the enterprise. But as the play progressed it proved to have been unimportant in comparison to the unspeakable crime of telling the truth about other people's mining propositions that were candidates for public money. The Scheftels corporation had laid it down as a set rule that an established reputation for accuracy of statement was a great asset for any promoter or broker to have. To gain such prestige the principle was followed in the nationwide publicity which emanated from the house that, no matter whom the truth hurt or favored, it must be told always, when publishing information regarding the value of any listed or unlisted security. Space in the Scheftels Market Letter or the news columns of the *Mining Financial News* was unpurchasable.

The enforcement of this rule was a wide departure from prevailing methods. But that didn't make us hesitate. Having felt the speculative pulse for years, I knew its throb. The public, after losing billions of dollars, were becoming educated. The rank and file of mining promoters—high and low—in Wall Street still believed that "one is born every minute and none dies." But I and my associates didn't. An uneducated public had been unmercifully trimmed in scores of enterprises backed by great and respected names. Speculators were ravenous for the truth. We decided to give it to them. We gave it to them straight.

This publicity system brought about the ruin of the Scheftels corporation through the powerful enemies it made. The policy was right all the same. Persisted in, nothing was or is better calculated to strengthen the demand for all descriptions of meritorious securities. The Scheftels corporation was the pioneer in the exploitation of this principle as a fundamental and underlying basis of brokerage and promotion. In pioneering this policy, however, the Scheftels company was sacrificed to the prejudices and wrath of the old school of promoters.

THE FIRING OF THE FIRST GUNS

BEFORE THE SCHEFTELS corporation was on the Street three months it almost came a cropper. On the strength of excellent mine news it purchased nearly 300,000 shares of Rawhide Coalition in the open market, up to 71 cents per share. A determined drive was made against the stock by mining stock brokerage firms which had sold it short. Bales of borrowed stock were thrown on the market by the crowd operating for the decline. The Scheftels company took it all in. Letters and telegrams were sent broadcast by market enemies urging stockholders to sell. A powerful clique had been losing big sums on the rise.

The Scheftels company published advertisements calling upon margin traders to demand delivery of their certificates. This expedient proved of small utility. The brokers continued to hold off deliveries to customers and sold and delivered to us all the stocks that they could borrow or lay hands on. The continued selling finally made inroads on the Scheftels corporation's cash reserve to a point that forced it one day to stand aside and leave the market to the sharpshooters. That day, in a few hours, approximately half a million shares of Rawhide Coalition changed hands out of a capitalization of 3,000,000 shares. The corporation's loans were called. This forced it to throw large blocks of stock on the market. A sharp break ensued. That was just what was wanted by the interests which were gunning for us. They covered their short sales at great profit.

In the midst of the mêlée the Scheftels company tendered a Stock Exchange house of great prominence, which had loaned it for the account of a Salt Lake firm of brokers $12,500 on 50,000 shares of Rawhide Coalition, the money to take up the loan. A representative of the Stock Exchange house sheepishly stated that his firm had loaned part of the pledged stock to out of town brokers. He asked for time. Under threat of dire consequences the Stock Exchange firm bought stock back from us in the open market that afternoon to supply the deficiency, and then made delivery of this stock back to us in lieu of that which they had parted with. It had been specifically stipulated by the Scheftels company when the loan was made that the certificates must be held intact and that the stock must not be loaned out or sold while the money loan was in force.

This experience was repeated frequently during the Scheftels career on the Curb. It cost B. H. Scheftels & Co. more than one million dollars, during the nineteen months of its existence, in giving loyal market support, in times of professional attack, to the stocks it had fathered or promoted and felt moral responsibility for.

Time and again the Scheftels company found among stocks delivered to it, against purchases made in the open market, the identical certificates it had pledged with loan brokers as collateral for loans, and which had been hypothecated by it with the specific proviso that the certificates were not to be used. It opened our eyes to one of the most commonplace practices, not only on the Curb, but also on the Stock Exchange. Hardly a failure occurs on any of the Exchanges or on the Curb that does not reveal customers' certificates, which were originally pledged with the understanding that they were not to be "used," in the strong boxes of others.

The first grievous offense of the publicity forces of the Scheftels corporation against Wall Street's "Oh let us alone" promotion combine was a wallop in April and May, 1909, through the Scheftels market literature, at Nevada-Utah.

The combination which owned control took with bad grace the strictures on the property. We heard an awful underground roar. At that time the price of Nevada-Utah stock was around $3. The Scheftels Market Letter said that there was probably not 30 cents of share value behind the property. The price immediately began to crumble. It has been tobogganing ever since. The stock at the beginning of September of this year was quoted at 37½ to 50 cents.

Such a thing as printing facts which would enlighten stockholders and the public as to the actual value and condition had not before been heard of when such enlightenment ran contrary to the plans of strongly entrenched promoters on the Street.

The campaign against Nevada-Utah, therefore, directed widespread attention to B. H. Scheftels & Co. and the *Mining Financial News*.

Following the Nevada-Utah disclosure, the Daily Market Letter and the Weekly Market Letter of the Scheftels corporation and the *Mining Financial News* took a good, strong, husky fall out of the La Rose Mines Co., capitalized for $7,500,000. The La Rose owns one of the greatest producing mines in the Cobalt silver camp. A market scheme was in progress, with La Rose as the medium, and W. B. Thompson, of Nipissing fame, as a chief manipulator. We called a halt to the game when the price reached a high of $8.50, and saved the public a huge sum of money. Under our campaigning the stock declined to $4, a decrease of $6,750,000 in the market value of the capitalization. This made W. B. Thompson and his associates the implacable enemies of the Scheftels company and myself. We didn't worry much. We were catering to the public. Indeed, we were pleased with our work.

Following this incident, the Scheftels Market Letter and the *Mining Financial News* took a smash at a mining stock deal in which W. B. Thompson and the Guggenheims were jointly interested. It was the now notorious Cumberland-Ely-Nevada Consolidated merger. Later the merger was enlarged and took in the Utah Copper Co., or rather the Utah Copper Co. took in the others, and the Scheftels propaganda found another opportunity to do a great service for the stockholders of Nevada Consolidated.

Our attack hurt the Guggenheim reputation among investors all over the country and contributed to reduce their influence over the large stockholding body—more than 6,000 men and women—of Nevada Consolidated. Though finally successful, the Guggenheims were sore from the lashing and exposures to which they had been subjected. As for the Scheftels company and the *Mining Financial News*, they had still further established the honesty and value of their publicity service.

A market scheme to balloon the price of Ray Central Copper Co. shares to several times their value was a precious enterprise against which we trained our publicity guns and fired several effective broadsides. The effort of the promoters to connect with the public purse here would not have been half so sensational if men of lesser prominence were identified with the operation. In our bear publicity on this one we minced no words. In doing so we again hit another powerful interest—the Lewisohns.

Later the exposure by the *Mining Financial News* and the Scheftels Market Letter of market manipulations of the Lewisohn-controlled Kerr Lake still further endeared the members of these two organizations to that powerful faction, and more closely cemented the ties of fellowship between the ruling powers.

Keystone Copper, another Lewisohn baby, was put through its courses on the Curb while Kerr Lake was being played in a stellar rôle. The deal in Keystone was an unobstrusive little thing, but awful good as far as it went from the one-sided point of view. I turned the searchlight of publicity on Keystone.

The Scheftels Market Letter and *Mining Financial News* disclosures in the interests of speculators and investors regarding Nevada-Utah, La Rose, Cumberland-Ely, Nevada Consolidated, Utah Copper, Ray Central, and Kerr Lake were sensational enough, but they by no means included all of the work in this line. During 1909 this publicity literature took in practically every important mining company whose shares were traded in on the New York Curb. The unpleasant truths these forces were obliged to tell from time to time touched the delicate sensibilities of many leading lights on the street. These had grown accustomed to an unvarying diet of sweets.

It would seem that their appetite for saccharine provender would have become cloyed and that a change would be a grateful relief. It was not. The truth was distasteful. It interfered with the noble industry of mining the public and it cut down the profits of that end of the game. In keeping up the record of day by day market and mine developments these publicity agents punctured many a rainbow-tinted balloon. Very frequently they gave to the public its first definite and intelligent idea of real value behind promotions and in properties. Where market prices represented an overplus of hopes and expectations the truth was told. The aim was to take mining speculation out of the clouds and plant its feet firmly on earth.

In this laudable effort we ran counter to the plans of the mighty. We also violated the vulgar unwritten rule of some of the Wall Street fraternity —never educate a sucker. Our publicity work caused a readjustment of judgment and market values, besides those already mentioned, on such stocks as First National, Butte & New York, Trinity Copper, Micmac, Ohio Copper, United Copper, Davis-Daly, Montgomery-Shoshone, Goldfield Consolidated, Combination Fraction, British Columbia, Granby, Cobalt Central, Chicago Subway, and sixty to eighty others.

The live wires of our publicity service blistered the flesh of the Guggenheims, the Thompsons and the Lewisohns, and perturbed their widely diffused affiliations, connections and allies, including John Hays Hammond, J. Parke Channing, and E. P. Earle; also Charles M. Schwab, E. C. Converse, B. M. Baruch, United States Senator George S. Nixon, George Wingfield, Hooley, Learned & Co., many other New York Stock Exchange houses, a group of powerful corporation law firms, a noted crowd of influential politicians, Curb stock brokers who had grown fat executing manipulative orders for the inside, bankers who carried on deposit the cash balances of the mining companies, and even J. P. Morgan & Co., who were partners of the Guggenheims in their Alaska ventures and were for a time said to be meditating a merger of the copper companies of the country with those controlled by the Guggenheims as a nucleus.

THE STORY OF ELY CENTRAL

BY KEEPING SPECULATORS out of stocks that were selling at inflated prices,

Before going to prison after his trial in 1911, George Graham Rice promoted the copper mines at Ely which boomed for almost two decades beginning in 1902. Long-line teams brought in lumber, machinery and other supplies before the coming of the railroad in the fall of 1906

the Scheftels corporation and the *Mining Financial News* became endeared to a great popular moneyed element. The public was saved huge sums of money.

This, however, only carried out the negative end of a grand idea. The affirmative demanded that the Scheftels corporation must put its followers into a stock or stocks where they could actually make money.

The Scheftels corporation was on the eager lookout for a genuinely high classed copper mining proposition. It found what it was looking for in Ely Central, a property that is sandwiched in between the very best ground of the Nevada Consolidated, is bordered by the Giroux and occupies a strategic position in the great Nevada copper camp of Ely, birthplace of what is probably the greatest lowest cost porphyry copper mine of America.

By invading the Ely territory as promoter and annexing Ely Central, the Scheftels corporation committed what was probably, to the interests among whom our publicity work had wrought greatest havoc, an unpardonable crime. We butted into the very heart of the game, and became a disturbing factor in their mining operations.

The Ely Central property consists of more than 490 acres. Years before, in the early days of the camp, it had been passed over by the geologists and promoters who selected the ground for the Nevada Consolidated, Giroux and Cumberland-Ely, because it was covered by a non-mineralized formation called rhyolite. As development work progressed and the enormous value of the surrounding mines was disclosed, it dawned on their owners that they might have made a mistake and that it would be just as well to obtain possession of the Ely Central property.

The ground was especially valuable to the Nevada Consolidated, if for no other reason, as mere acreage to connect up and make compact the properties owned by them. The second demonstration of their bad judgment was the fact that, having planned to mine the Copper Flat ore-body by the steam shovel method, they overlooked the value of the Ely Central property as affording them the only practical means of access to the lower levels of that pit for operation by the steam shovels.

Investigation had disclosed to me that the evidence which had been adduced by mine developments on neighboring properties was all in favor of copper ore underlying the Ely Central area. The rhyolite, which covered Ely Central, was a flow, covering the ore, and not a dyke, coming up from below and cutting it off.

Why was the property idle? Inquiry revealed that the Ely Central Copper Co. was $89,000 in debt, and that a pre-panic effort to finance the corporation for deep mine development had failed. The panic of 1907-8 had crimped the promoters and they could not go ahead.

The Scheftels corporation entered into negotiations with the Pheby brothers and O. A. Turner, who held the control, for all the stock of the Ely Central company that was owned by them. During the progress of negotiations, early in July 1909, I heard that the Guggenheims and W. B. Thompson were very much put out to learn that the Scheftels company was about to finance the company. They had belittled the value of the property, as would-be buyers are prone to do the world over.

Before I entered upon the scene the Pheby brothers had found themselves objects of persistent and mysterious attacks. Their credit was assailed in every quarter and they found themselves ambushed and bushwhacked in every move they made. They were forced into a position where it was believed they would accept anything that might be offered them for their interest in Ely Central. As fate would have it, the Scheftels company entered the race at this psychological moment.

Summed up, the Scheftels company actually contracted for 1,280,571 shares out of 1,600,000, which represented the increased capitalization for a total sum of $1,158,916 or at an average price of 90½ cents per share. The time allowed for the payment of all the money was nine months, stipulated payments being agreed upon at regular intervals in between. The immediate effect of the arrangement was this: A dormant property, in debt and lying fallow, was metamorphosed into a going concern with good prospects of soon becoming a proved great copper mine, with an assured income to defray the expenses of deep mine development on a large scale, and a market career ahead of it that might be expected to match any that had preceded it in the Ely district from the standpoint of public interest.

During the progress of the negotiations the stock sold up to $1 per share. The selling for Philadelphia account of a large block of stock in the open market dropped the price back, of a sudden, to 50 cents. The Scheftels company bought stock on this break and urged its customers to do likewise. On the day the deal was concluded the market had rallied to 75 cents. Fully six weeks before the deal was arranged the Scheftels Market Letter and the *Mining Financial News* had begun to urge the purchase of the stock. The Scheftels organization was not hoggish. The establishment was willing that the public should get in on the cellar floor. There were nearly 300,000 shares outstanding, which the Scheftels corporation had not corralled in its contract.

Readers of the Market Letter and the *Mining Financial News* fell over one another to get in on the good thing. Therein they were wise. By early

September the price had advanced in the market to $1. The Scheftels publicity was strong in favor of the stock. But it had not yet put on full steam. It was waiting for an engineer's report to make doubly sure it was right.

Col. Wm. A. Farish, a mining engineer of many years' experience and a man with a high reputation throughout the whole of the Western mining country, had been sent by the Scheftels company to make a report on the Ely Central. Years before Colonel Farish had reported on the Nevada Consolidated properties and outlined the very methods now being used for recovering its ores. But Colonel Farish had been ahead of his time, and the capitalists in whose interest he was acting were not prepared for such a radical step in advance of the then-existing methods, nor to believe that copper ores of such low grade could be mined at a profit, especially 140 miles from the nearest railroad. Times and conditions changed, and the 140 miles were spanned by a well-equipped rail connection.

Colonel Farish's opinion verified our fondest expectations. The report set forth that the mine possibilities of Ely Central were nearly as great as those of the Nevada Consolidated itself. On the basis of this report, which was made in September, the project acquired a new significance. Development operations were undertaken to prove up the ground in an endeavor to demonstrate the existence of the 33,000,000 tons of commerical porphyry ore which Colonel Farish indicated in his report would likely be found within the boundaries of the southern part of the Ely Central property.

The prospect fairly took the Scheftels organization off its feet. We were dazzled. We saw ourselves at the head of a mine worth $25,000,000 to $40,000,000. No time was lost in organizing a campaign to finance the whole deal. Having no syndicated multi-millionaires to back it up, the Scheftels corporation went to the public for the money, the same as hundreds of other notable and successful promoters had done. The ensuing publicity campaign to raise capital has been described in hundreds of columns of newspaper space as one of the most spectacular ever attempted in Wall Street.

I had absolute faith in the great merits of Ely Central, a faith that has not been dimmed in the slightest degree by the vicissitudes through which the company, the Scheftels corporation, and myself personally have passed. Within thirteen months the Scheftels corporation caused to be spent for mine development more than $150,000, and on mine and company administration an additional $75,000. When the Scheftels company was raided by the government on September 29, 1910, and a stop put to further work the expenses at the mine had averaged for the nine months of that year above $15,000 a month. Work was going on night and day. Every possible effort was being made to prove-up the property in short order. Core drills sent down from the surface had already revealed the presence of ore at depth, and I am sublimely certain that another month or two would have put the underground air drills into contact with a vast ore body identical in quality and value with that lying on either side in Nevada Consolidated acreage.

Ely Central was the New York Curb sensation in 1909-1910. I used the publicity forces which had been so successful in protecting the public against the rapacity of multi-millionaire mining wolves to educate them up to the speculative possibilities of Ely Central.

Up went the price. Between the first of September and the middle of October the market advanced to $2⅜. On October 13th advices reached us that 30 per cent copper ore had been struck in the Monarch shaft. The Monarch is an independent working, far removed from the area that is sandwiched in between the main ore bodies of the Nevada Consolidated. We were highly elated. The prospect looked exceedingly bright to us, and there was no longer any hesitation in strongly advising our following to take advantage of an unusually attractive speculative opening.

The market boomed along in a most satisfactory way. By October 26th the price reached $3; on November 3rd it was $4 a share, and three days thereafter $4¼ was paid.

The expenses of the Scheftels company on publicity work at this time amounted to about $1,000 a day. Money for mine development on Ely Central was being spent as fast as it could be employed. We were trying to sell enough stock at a profit over the option price to defray the publicity expenses, keep the mine financed, and meet our payments on the option, but no more. We were not making any effort to liquidate on a large scale, a fact which was reflected in the advancing quotations.

When the price of Ely Central hit $4 in the market, the Scheftels company rated itself as worth from $3,000,000 to $4,000,000. I had visions of leading the Guggenheims and Lewisohns and Thompsons up the Great White Way with rings in their noses. Nat C. Goodwin, who had a 25 per cent interest in the Scheftels enterprise, enjoyed similar visions, only his fancy ran to building new theaters for all star casts.

While Ely Central stock was going skyward and all the speculating world was making money in it, our publicity forces were busily driving the bald facts home regarding La Rose, Cumberland-Ely, Nevada-Utah and other pets of the mighty. Our batteries never let up for a moment. These various attacked interests were getting ready to strike back. If their movements had been directed by an individual general they could not have worked with more community of interest. One day the sky fell in on us. The plans had been beautifully laid for our complete ruin. That we escaped utter annihilation was almost a miracle.

On Wednesday, November 3rd, the result of our market operations on the New York Curb was that we quit long on the day nearly 8,000 shares of Ely Central at an average price of $4. On the same day our customers ordered the purchase of nearly twice as much stock as they ordered sold. This indicated to us that the Curb selling was professional. There was nothing very remarkable about this performance because brokers doing business on the Curb very frequently play the market for a fall.

On Thursday, the day following, the Scheftels company was again compelled to purchase stock on the Curb in excess of sales to the extent of 7,600 shares, while on the same day the buy orders of house customers exceeded their orders to see at least three to one. The professional selling was now accompanied by rumors on the Curb which spread like the smell of fire that trouble of some dire sort was pending for the Scheftels company. Most of this emanated from an embittered brokerage quarter and we paid little attention to it.

On the succeeding day, Friday, November 5th, the professional selling was quieted to a point that compelled the Scheftels company to go long of only 6,600 shares on the day in its Curb market operations. The purchase of so small a block of stock excited no suspicion in the Scheftels camp, although it should have, because Scheftels' customers on this day purchased more than four times as much stock as they ordered sold, pointing conclusively to a great public demand and much shorting by professionals.

Then came the *coup de main*.

THE ASSAULT ON ELY CENTRAL

THE 6TH DAY OF November fell on a Saturday. The New York *Sun* of that morning published under a scare head a vicious attack on the Ely Central promotion. The attack was based on an article which was credited in advance to the *Engineering & Mining Journal* and appeared in the *Sun* ahead of its publication in that weekly. The *Sun* had been furnished with advance proofs. The Ely Central project was stamped as a rank swindle. Everybody identified with it was raked over and I, particularly, was pictured as an unprincipled and dangerous character, entirely unworthy of confidence and at the moment engaged in plucking the public of hundreds of thousands. It was stated that the Ely Central property had been explored in the early days of the Ely camp and found no value whatsoever from a mining standpoint. The

Scheftels corporation was accused of setting out in a cold-blooded way to swindle investors on a bunco proposition.

I was in my apartment at the Hotel Marie Antoinette at 9 A.M. when I read the *Sun's* story. The Scheftels company had thrown $85,000 behind the market during the three preceding market days to hold it against the attack of professionals.

I called the Scheftels office on the 'phone and gave instructions that a certified check for $40,000 be sent to Wasserman Brothers, members of the New York Stock Exchange, with orders to purchase 10,000 shares of Ely Central at $4⅛, which was the quotation at the close on the afternoon before. Orders to buy 15,000 shares more at the same figure were distributed among other brokers. The single order was given to Wasserman Brothers because I thought it good strategy. They are a house of undoubted great responsibility and it seemed to me that their presence in the market on the buying side would have an excellent tonic effect.

During the two hours' session I held the 'phone, receiving five minute reports from the scene of action. Mr. Goodwin was at my side. At ten minutes to twelve the brokers had reported the purchase, on balance, of 24,225 shares. Had they purchased 675 shares more they would have completed the orders that were outstanding and it would have been up to me to decide whether to lend further support or not. By that time my figures showed that the Scheftels corporation had thrown behind the market $200,000 in four days to hold it and I was beginning to have that funny feeling. During the last few minutes of the Saturday Curb session the selling ceased and it seemed that possibly my fears were unfounded.

On Sunday, the 7th, my hopes went a-glimmering. All the New York papers featured scathing articles, using as authority the *Engineering & Mining Journal's* attack, which had appeared on the previous afternoon. Dispatches indicated, too, that the papers of Boston, Chicago, Los Angeles and San Francisco had played it up on the front page as the most shocking mining stock scandal of the century.

By Monday, the whole country had been plastered with the sensation. Of course my early past, all of which was a family affair and had transpired fourteen years prior, long before I essayed to enter the mining promotion field, was dragged out of the skeleton closet. It lent verisimilitude to the stories.

After reading the Sunday newspapers, I grasped the meaning of the move and marshalled our forces. It was plain that we had been marked for the sacrifice. It looked as though we hadn't a chance in a million of weathering the onslaught if we lent the market further support. There were about 500,000 shares of Ely Central in the public's hands, and, without close to $2,000,000 in ready cash to throw behind the market, we could not be certain of staying the tide. We didn't have anything like that sum. Personally I did not give up the fight, but the outlook was mighty blue.

All day Sunday trusted clerks of the Scheftels company worked on the books, making a statement of the stop-loss orders and good-till-cancelled orders of customers. On Monday morning the newspapers contained aftermath stories of the *Engineering & Mining Journal's* arraignment. The air was surcharged with the impending calamity.

THE CLASH OF BATTLE

WITH A LINE OF defense carefully outlined, I approached the fray. First, the Scheftels corporation placed with reliable brokers written orders to sell at the opening the stocks that were specified in the stop-loss and good-till-cancelled orders of customers. Not an order to sell a share of inside stock was given. It was also decided not to place any supporting orders until after the market opened and it could be determined with some degree of accuracy what the volume of stock amounted to that was pressing for sale.

Just before the market opened I could see from my office window a dense crowd of brokers assembled around the Ely Central specialists. Although ominously silent, they were struggling for position and were tensely nervous. It was plain that the over Sunday anti-Scheftels newspaper publicity had racked Ely Central stockholders and created a panicky movement to liquidate, which was about to find vent in violent explosion. It was evident that the Scheftels corporation would have to conserve every resource if the day was to be saved.

The market opened. Instantly there was terrific action. Hundreds of hands were waving wildly in the air. Everybody wanted to sell and nobody wanted to buy. The chorus was deafening. Screams rent the air. The tumult was heard blocks away. Every newspaper had a man on the spot. Brokers from the nearby New York Stock Exchange left their post and came to see the big show; the Stock Exchange was half emptied. The spectacle had been advertised widely and everybody was keenly awake and wrought up to a high pitch of excitement over what had been scheduled to occur.

Had the Scheftels brokers been supplied with orders to buy one-quarter of a million shares of stock at the closing market price of the Saturday before, $4⅛, it was very apparent that they would have been unable to hold the market. The opening sale was at $4. Downward to the $3 point the stock traveled, breaking from 25 to 50 cents between sales. Through $3 and on down to the $2 point the price crashed. Blocks of 10,000 shares were madly thrown into the vortex of trading. The Curb was a struggling, screaming, maddened throng of brokers. Every trader appeared to be determined to crush the market structure. At $2 a share there was a temporary check in the decline, but the bears renewed their onslaught, gaining confidence by the outpour of selling orders. Within less than an hour after the opening the stock hit $1½ a share. At this juncture the Scheftels broker in Ely Central reported that he had executed all the stop-loss and good-till-cancelled orders entrusted to him with the exception of 19,000 shares.

"The Scheftels company will take the lot at $1½," I said.

In lending succor at $1½ per share I was really stretching a point, although at this figure the next market shrinkage of the Ely Central capitalization was in excess of $3,000,000. This melting of market value was awful to contemplate. On the other hand the newspaper agitation was unmitigated in its violence, stockholders were convulsed, a break of serious proportions was certain, and it was up to me to conserve every dollar. The moment the Scheftels bid of $1.50 a share made its appearance on the Curb and the selling from the same source for the account of customers was discontinued, it was seen that the force of the drive had spent itself, at least for the time being. Support now came from the shorts. They started to cash their profits on their short sales of the days previous. Crazed selling was transformed to frantic buying.

The scene at this juncture was dramatic. It was the momentary culmination of a cumulative, convulsive cataclysm. In refraining from selling for its own account the Scheftels corporation violated one of the sacred rules and privileges not only of the New York Curb but of the New York Stock Exchange. In both of these markets it is the custom, where brokers have advance information of an impending calamity, to beat the public to the market and get out their own lines first, leaving customers to take care of themselves.

By deftly feeding stock to bargain hunters and to the shorts at intervals and buying stocks when it pressed for sale from frightened holders at other periods the Scheftels company was able to support the market that afternoon to a close with sales recorded at $2 a share. The cash loss of the Scheftels company on its Curb transaction in Ely Central that day was $60,000. This fresh sacrifice was needed to steady the market.

Tuesday, the following day, the daily newspapers belched forth further tirades of abuse and calumny. The market crash in Ely Central was held up to the public as proof positive that the project was a daring swindle. The raid on the stock in the market was renewed. A Johnstown flood of liquidation ensued. Fluctuations were violent. Opening at $2, the price was forced down to $1. It afterward rebounded to $2, but the waters would not subside, the stock was hammered again and it closed at $1 per share. To meet the oncoming emergencies the Scheftels corporation was obliged to fortify its cash reserve in the only one way that offered. It was compelled to convert a large part of its reserves of securities into cash and it had to sell on a declining market. Many accounts were withdrawn by timid customers, and the Scheftels company was further called upon to give stability to Rawhide Coalition and Bovard Consolidated, other stocks which it had been sponsoring in the markets. Loans were called by brokers with whom the Scheftels company was carrying stocks, deliveries were frantically tendered to the Scheftels company of stocks it had purchased at previous high levels, and a financial onslaught made generally that would undoubtedly have sunk the Scheftels' ship but for the fact that we had backed up in the nick of time, had measured our distance, had gone just so far and not too far, and had kept on the firing line.

An exceedingly gratifying feature of the sensational day was the way in which our friends stood by us. The venom and selfishness of the overwhelming assaults that had been made upon us convinced many of the public that we were being made the victims of a special attack, and with the natural impulse that governs honorable men they gave testimony to their confidence in us.

On Wednesday the campaign terminated. Ely Central weakened an eighth from the $1 point, the closing of the day before, recovered to sales at $1¾ and closed at $1½ bid, $1⅝ asked.

All day long our offices were thronged with newspaper reporters and with palefaced, agitated customers. Our clients felt their helplessness in such a tumult of warring forces. The only thing they could do was to stand by and watch developments as the battle waged. It was a proud moment for me when, at the end of the day's market, I mounted the platform in the Scheftels customers' trading room, gave voice to a shrill cheer of triumph and wrote on the blackboard the following:

"We have not closed out a single margin account! We are carrying everybody!"

The scene which followed warmed the cockles of my heart. I was literally mobbed, but it was a friendly mob. We all joined in a season of noisy rejoicing. That we should have been able to survive the three days' siege with minimum losses to customers and without sacrificing a single margin account was a signal achievement. I doubt if there are many cases like it in the history of Wall Street.

Scores of telegrams were received from out of town customers to whom the margin respite was wired. One of these read:

> You may look for a tidal wave of business. Your princely action warrants 21 guns for the House of Scheftels.

Another one was to this effect:

> The whole situation was greased for your descent. It was a shoot the chutes and a bump the bumps proposition. Congratulations on your survival.

Hundreds of letters of a similar tenor poured in upon us. Many of these came from the camp of Ely itself, where large blocks of the stock were held by mining men on the ground.

Thursday the stock closed at $1¾; Friday it advanced to sales at $1⅞ and hung there.

The Scheftels organization now drew its first long breath. Friends and enemies alike marveled how the corporation had managed to survive. We had held the fort, but at murderous cost.

I got busy with the publicity forces at my command. Through the Scheftels Market Letter and the *Mining Financial News* the story was told of the whole dastardly campaign.

The Weekly Market Letter of the Scheftels company on November 13, 1909, devoted 24 columns to the story of the raid.

That the Guggenheim-managed Nevada Consolidated was well pleased with the publication of the *Engineering & Mining Journal's* attack seemed clear to me. The reason was this: In its attack the *Engineering & Mining Journal* stated that two drillholes put down by the Nevada Consolidated in the immediate vicinity of Ely Central had failed to show better than nine-tenths of one per cent copper ore which, the article said, was below commercial grade. (At this late date, October 1911, they are mining ore in the steam shovel pit of the Nevada Consolidated that will not average more than eight-tenths of one per cent copper transporting it to the concentrator, more than twenty miles away, and treating it at a profit. But this is not the point.) An engineer of international prominence telegraphed the Scheftels company from Ely as follows:

> Two drill holes mentioned in *Engineering & Mining Journal* article were completed only last week. Results must have been telegraphed to New York.

> These holes gave great trouble on account of caving ground. I heard drill runners say they were stopped on that account and were in ore in bottom. In any case, it is not conclusive of unpayable ore in vicinity. This condition often occurs.

I could write a book in reply to the *Engineering & Mining Journal's* tirade, showing the utter flimsiness of the statements it made. Limited space forbids anything more than an outline.

Charles S. Herzig was employed to report confidentially on the property. Mr. Herzig's report was later checked up by Dr. Walter Harvey Weed, a great copper geologist of known high standing who was formerly one of the principal experts of the United States Geological Survey and was himself a frequent contributor to the *Engineering & Mining Journal*. Dr. Walter Harvey Weed wired to the C. L. Constant Co., the metallurgists and mining engineers, from Ely, as follows:

> After making a most thorough examination my opinion is Southern part Ely Central property is covered by rhyolite capping. Geological evidence demonstrates that the porphyry extends eastward (through Ely Central) from steam shovel pit and with excellent chance of containing commerical ore beneath a leached zone. A well defined strong fault separates steam shovel ore from rhyolite area and this fault plane may carry copper glance (very rich copper ore) or recent origin, due to descending solutions. The iron-stained jasperoid croppings in the limestone areas give promise of making ore in depth on Ely Central property as they do in Giroux.

The *Engineering & Mining Journal* said in its article that the northern portion of Ely Central showed the Arcturus limestone of the district. It stated that in this limestone at various places there is a little mineralization but never during the history of the district were any profitable results obtained. As against this, Engineers Farish, Herzig and Weed reported that the limestone areas on Ely Central would likely show the presence of mines. As a matter of fact, Giroux, neighbor of Ely Central, had sunk through this limestone and opened one of the richest bodies of copper ore ever disclosed.

The *Engineering & Mining Journal* said that in representing that pay ore is likely to exist in the area of Ely Central sandwiched in between the two big mines of Nevada Consolidated, the Scheftels company was practicing deception. Not only did Messrs. Farish, Herzig and Weed report in favor of the likelihood, but it is now a commonly accepted fact that, unless all known geological indications are deceptive, Ely Central has the ore in this stretch of territory. A report made as late as September 1911, by engineer Richard T. Pierce, for the reorganization committee of Ely Central, expresses the opinion that an area 1,300 feet by 1,900 feet at the southeast end of the Eureka workings "will be found to contain mineralized porphyry, with reasonable assurance that commerical ore will be had in it."

Mr. Herzig's first telegram from Ely after examining the Ely Central property was to this effect:

> There is no question that the rhyolite was deposited in Ely Central after the enrichment of the porphyry. The fault that limits the rhyolite in the Nevada Consolidated pit is indicated by several feet thickness of crushed mineralized porphyry-rhyolite ore, which is a positive evidence that the porphyry was enriched before the faulting. The limestone and contact areas owned by the company, in my opinion, have great potential value. The indications are in every way similar to Bisbee. Rich carbonate ore has been encountered on the Clipper and Monarch claims of Ely Central and I look forward to seeing big ore bodies opened up at these places.

Reports of both these engineers, many thousand words in length, made later, confirm these messages.

What probably convinced me more than anything else of the inaccuracy of the statement regarding the Ely Central property by the *Engineering & Mining Journal* was the attitude of Charles S. Herzig. He is my brother.

Up to within thirty days of the appearance of the attack in the *Engineering & Mining Journal* I had not set eyes on him in fifteen years. A graduate of the Columbia School of Mines, he had in the interim examined mining properties in South Africa, Egypt, Australia, the East Indies, Siberia, every European country, Canada, Mexico, Central America, South America and the United States in the interests of some of the world's greatest financiers. These expert examinations had covered deposits of gold, silver, copper, lead, zinc, coal and other minerals. In the engineering profession he is known as an expert who has his first failure yet to record. His standing is unquestioned as an engineer and a mine valuer.

I had heard some criticism of the Farish report, made by engineers of the modern school, in which it was pointed out that Colonel Farish had failed to give scientific reasons for all of his deductions. I asked Captain W. Murdoch Wiley, then a member of the C. L. Constant Co., assayers, metallurgists and mining engineers, whether he could induce my brother to make an examination. I did not approach Charles myself, because we had been estranged. So it was that when he returned from Europe after an absence of many years, he had not even looked me up. Captain Wiley arranged for a meeting at the Engineers' Club. I went there, and was formally introduced by Captain Wiley to my brother across a table.

"What will you take to make a report on Ely Central?" I asked in the same matter of fact way I would have addressed a stranger.

"What's the purpose of the report?"

"The Scheftels company wants confidential, expert information such as you are qualified to give as to the value and prospects of the property," I answered.

"I'll take $5,000," he said, "but only on one condition. I am going to the Ely and Ray districts to report for English capitalists, and I can take your property in at the same time. My report is not to be published and I reserve the right to make a verbal instead of a written one. If you really want to know what I think of the property, I am quite willing to give it a careful examination and let you know. Because of the stock market campaign you are making, I would not accept your offer if, did I report favorably, your idea would be to make use of the report in the market."

The bargain was struck. A few days later Mr. Herzig received $2,500 from the Scheftels company, on account, and a check for traveling expenses. He left for Ely.

On the Saturday morning when the New York *Sun* article appeared containing the excerpts from the *Engineering & Mining Journal's* onslaught, I wired my brother substantially as follows:

> Savage attack in *Engineering & Mining Journal* on Ely Central. If your report on property is favorable, I beg you to let us have it by wire and allow the use of it to counteract.

An hour later I followed it up with another message telling him not to wire any report. I set forth that because he was my brother, it might prove of little avail, now that the publication had been made, and that it might only tend to do him personal damage in the profession because of the unqualified manner in which the *Engineering & Mining Journal* had taken a stand against the property. In reply he wired Captain W. Murdoch Wiley the short but decisive report already quoted herein, regarding the geological reasons why Ely Central should have the ore, which afterward was fully verified by Dr. Walter Harvey Weed in the message also reproduced in the foregoing. In a letter from Ely to Captain Wiley confirming the message, the original of which is in my possession, Mr. Herzig said:

> I have formed a very favorable opinion of the property. I feel that it has the making of a big mine, and under the circumstances I am willing to stand a little racket for a time.

The same day he wired Captain Wiley to buy for his account 2,500 shares of Ely Central at the market price, which order was executed through the Scheftels company.

Editor Ingalls of the *Engineering & Mining Journal* and my brother had been friends for years. My brother had been employed early in his career by the Lewisohns, Guggenheims and the Anaconda Copper Co., and later in Europe, Australia and India by mine operators of even higher class. Up to the time when the *Engineering & Mining Journal's* attack appeared he had not committed himself on Ely Central. When he did commit himself it was with the foreknowledge that in doing the unselfish and courageous thing his name would be besmirched if under development Ely Central turned out to be what the *Engineering & Mining Journal* had declared probable. In that event his relationship with me would be held up as positive proof of duplicity and it would look bad for him. The fact that under all these circumstances he jumped into the breach satisfied me that the attack of the *Engineering & Mining Journal* was unjustified.

A BOMBSHELL IN THE ENEMY'S CAMP

As soon as the Scheftels corporation was able to obtain a copy of the corroborative report of Dr. Walter Harvey Weed, which the great copper geologist made to the C. L. Constant Co., it filed a libel suit against the *Engineering & Mining Journal* for $750,000 damages. Simultaneously Mr. Scheftels filed another suit for an additional $100,000 in his own behalf.

The filing of the Scheftels libel suits against the *Engineering & Mining Journal* was a bombshell. It was formal notice to the forces arrayed against us that we did not propose to be made victims of an unholy hostility and that we were determined to proceed along old lines and not abate in the slightest our wide-open publicity measures. It was also noticed that we proposed to go through with the Ely Central deal.

After it became evident that we intended to keep on fighting, the Scheftels offices were openly visited and inspected in detail one day by the late police inspector McCafferty. In a bullying manner this police official let it be known that we were in official disfavor with him. His manner could hardly have been more offensive if he had been invading a den of counterfeiters. Mr. McCafferty did not specify just what he was after or just what he expected to find, but he made it plain to us that we were marked and that he had it in for us. He stalked scowlingly through the entire establishment and made vague threats of what was in store for us.

Late that night I learned that the inspector had invaded the living rooms of my associate, Nat C. Goodwin, where he delivered himself somewhat as follows:

"What are you fellows trying to do, anyway? What are you trying to put across on us? Do you think we are going to stand for any such newspaper notoriety as you are getting and watch it with our arms folded? Do you think we are fools or crazy, or what? I want you to understand that you fellows have got your nerve with you. Get busy or the police will be on your backs tomorrow!"

I told Mr. Goodwin that our enemies had evidently sicked the inspector on to us, but that I didn't think any action would be taken. We were victims and not culprits, and unless, indeed, the United States was Russia, nothing untoward could happen.

I promised Mr. Goodwin, however, that I would attend to the matter without delay. I laid all of the facts regarding the newspaper attack before a prominent citizen who promised forthwith to convey the information in person to the inspector or one of his superiors. He did so. That was the last we heard of the matter.

The *Engineering & Mining Journal*'s lawyers addressed themselves to customers of the Scheftels company, who had lost money in the market break in Ely Central from $4 to $1.50. By letter they urged them to send on a full statement of facts and suggested that they might be of service, and without charge.

Letters of this character were sent to large numbers of our customers, many of whom simply sent them to us. In some cases, however, customers who had read the attack in the *Engineering & Mining Journal* or quotations from it in widely circulated daily newspapers, needed but the letter from the lawyers to induce them to come forward with a complaint.

On the whole, this fishing expedition must have been something of a water haul and a disappointment, for the attorneys of the *Engineering & Mining Journal*.

The Post Office Department at New York, in January and February, sent letters broadcast to readers of the Scheftels Weekly Market Letter, asking whether the business carried on was satisfactory—the usual form that is used where a firm is under investigation. Scores of these letters were forwarded to us by customers with remarks to the effect that evidently somebody was after us. An inquiry of this sort is calculated to do terrible damage to the reputation and standing of any house that does a quasi-banking business. Our attorneys complained to inspector Mayer of the New York division of the Post Office that an injustice was being done. No more letters of the character described were sent out, because the early replies that were received by the inspector to his circular letter brought forth no serious complaints. However, it was afterward disclosed that the investigation did not cease here and that the Post Office Department continued to conduct a searching inquiry only finally to abandon its enterprise.

Enters upon the scene an associate of the *Engineering & Mining Journal's* lawyers defending that publication against our suit for libel. He called at the Scheftels offices and demanded from Mr. Scheftels information with regard to the account of C. H. Slack of Chicago. He got it. It showed that Mr. Slack had purchased 50,000 shares of Bovard Consolidated at 10 cents per share, for which he had paid cash, and that Mr. Slack had purchased an additional 100,000 shares at 14¼ and 14¾ cents per share; and that Mr. Slack had refused, after the market declined to below the purchase price, to pay the balance due, because of delayed delivery.

The delay in delivery was accidental. The Scheftels company actually had in its possession two million shares of the stock or more, and the delivery would have been tendered earlier but for the fact that the raid on Ely Central had piled up so much work for the clerical force that everything was set back. We knew of no legitimate excuse for Mr. Slack, because he could have ordered the stock sold at any time, delivery or no delivery. The Slack transaction receives amplification here, because later, when the Scheftels corporation was raided by a special agent of the Department of Justice, it figured as one of the cases cited by the agent in the warrant sworn to by him against B. H. Scheftels & Co. as proving the commission of crime.

Another case about which Mr. Scheftels was asked to give full information was that of D. J. Szymanski, a corn doctor at 25 Broad Street. Mr. Scheftels had urged the doctor to buy Ely Central when it was selling at 75 cents before the rise. Later, when the advance was well under way, above the $3 point, the doctor bought some stock through the Scheftels corporation. When the price hit $4 he was urged to take profits. He refused to do so. When the attack began and the price broke badly the doctor saw a big loss ahead.

He called at the Scheftels' office and begged for the return of the money he had lost in his Ely Central speculation.

The investigation was heralded among the brokers and caused much market pressure on the stocks fathered by the Scheftels company. We were not dismayed. To strengthen our position and to give added token of our good faith we increased our development operations at the mines. Our expenses in that quarter were swelled to the limit of working capacity on our underground explorations, as I realized that our salvation might depend on making good in quick order with Ely Central from a mining standpoint. We knew the ore was there and that it was up to us to get it before our enemies got us.

A GOVERNMENT RAID IS RUMORED

OUT OF A BLUE SKY late in the month of June came news to the Scheftels office that a newspaper reporter on the New York *American* had stated that he had seen a memorandum in the city editor's assignment book to watch out for a Scheftels raid by the United States government. The information was reliable and it gave us a shock. Yet the thought that the powers of a great government like the United States could be used to crush us without giving us a hearing seemed unbelievable.

To be on the safe side Mr. Scheftels, accompanied by an attorney of high standing, visited Washington. They went direct to the Department of Justice, where Attorney-General Wickersham's private secretary, after a friendly conversation, referred them to the chief clerk. He reported, after a search of twenty five minutes' duration, that there was no charge against B. H. Scheftels & Co. He even volunteered the information that he did not know that such a firm was in existence.

It afterward developed that at the very time Mr. Scheftels and the attorney were at the Department of Justice a special rubber shoe investigation was on under the dual direction of a young Washington lawyer on Attorney-General Wickersham's personal staff, and a Special Agent of the Department of Justice. The latter had been given extraordinary powers as a special agent of the Department of Justice, ostensibly to clean out Wall Street.

Satisfied they were in the wrong place, Mr. Scheftels and the counsellor departed from the Attorney-General's office for the Post Office Department. They were referred to Chief Inspector Sharp. The lawyer requested that the Scheftels corporation be given a hearing before any action was taken on any complaints that might reach the department. Mr. Sharp agreed to this on condition that the attorney would agree for the Scheftels company that an inspection of the books of the corporation would be permitted on demand at any time. There was a ready assent. A memorandum to this effect was left with Inspector Sharp.

Mr. Scheftels left the Department with positive assurance that no snap judgment would be taken. Edmund R. Dodge of Nevada, personal counsel of B. H. Scheftels & Co., then addressed a letter to U. S. Senator Newlands with the request that he take the matter up direct with the Postmaster-General.

Senator Newlands, under date of July 2, wrote Mr. Dodge that he had addressed a letter to the Postmaster-General with the request that notice be given to Mr. Dodge in case any complaint or information was lodged against the Scheftels corporation. A few days later Senator Newlands sent Mr. Dodge a letter from Theodore Ingalls, Acting Chief Inspector of the Post Office Department, in which Mr. Ingalls said it was the practice of the Department in case of alleged use of the mails for fraudulent purposes to give individuals against whom complaint has been made full opportunity to be heard either through person or counsel, should adverse action be contemplated as a result of the investigation of such allegation.

Feeling that our house had been securely safeguarded against surprise parties, I at this junction took a trip to Nevada, where urgent business matters required my attention. While I was in the West telegrams were sent me that the premier mail order mining stock bucketshop firm on Broad Street was flooding the mail and burdening the telegraph wires with urgent appeals to stockholders of Rawhide Coalition, one of our specialties, to sell out their holdings, as a severe break in the price of the shares was impending. Forewarned of this attack, I telegraphed instructions from Reno to meet the onslaught with a notice in the *Mining Financial News* addressed to investors, telling them to be on their guard.

My trip to the West made a pocketful of money for investors by my purchase of the control in the Jumbo Extension company on a monthly payment plan. The price of the stock tripled in the market. My re-entrance into the Goldfield camp was especially distasteful to the Nixon-Wingfield interests. Before I left Goldfield I was actually warned that the vengeance wreaked on the Sullivan Trust Co. would be visited on the Scheftels company for daring to reinvade the Goldfield district.

Late in August the Scheftels company endured what was probably the most severe strain it had been put to since its incorporation. We had been making heroic efforts to rally the price of our specialties on the New York Curb market. We were meeting with unusual resistance from professional sources. At the period of which I narrate, Ely Central had registered a low quotation of 62½ cents and we had successfully strengthened it to around $1. All the way up we met with heavy sales. One day deliveries crowded in so fast that three cashiers working in the cage were unable to keep up with the transactions. The business of the corporation had been heavy in the general list as well as in the house specialties. There was more than sufficient money on hand to finance all the transactions that day, but not, however, unless deposits were made in bank as rapidly as our own deliveries were made and collected for by our messengers.

About 2 o'clock in the afternoon a report reached the Curb that the bank checks of B. H. Scheftels & Co. were not being promptly certified. As this rumor gained currency the excitement on the Curb increased. The Curb concluded that we were at last busted. Motley throngs began to assemble in front of the offices. The fierce yells of brokers could be heard bidding for and offering Scheftels checks below their face value. A throng of the riff-raff of the Street swarmed in front of the building.

One or two individuals, implacable enemies who had repeatedly led the market onslaught against the Scheftels stocks, offered Scheftels checks for small sums at as low as 50 cents on the dollar. These were licked up by our friends who had been assured that we were financially all right and that some mistake must have been made at the bank.

Investigation showed that dilatory message service was responsible for the bank's delay in certifying. Our deposits did not reach the bank as promptly as they should have. As a special favor to us that afternoon, while the tumult in front of our doors was greatest, the bank continued to certify checks until 3:30 o'clock, extending the closing time 30 minutes. Then they reported that a comfortable cash balance was still on hand.

The next morning the newspapers started a jamboree. First page, last column, double-leaded, scare-head stories greeted every New Yorker for breakfast, telling him about the panic among Curb brokers to sell the Scheftels checks the afternoon before. Needless to say it was the kind of notoriety that was likely to do greatest injury to the House of Scheftels. If anything half as bad had been printed about the strongest bank in New York, that bank would have been forced to close its doors before the day was half over.

Nor did I for a moment underrate the danger of our position. Between two suns I managed to assemble $50,000 in addition to our cash reserves, with promises of as much more as was needed. We easily held the fort. At the end of the day's business I created a diversion by appearing in the Scheftels board room, flourishing a handful of $1,000 bills before the newspaper men. The scribes found the Scheftels corporation meeting all demands, and, at the end of the session, with a small bale of undeposited money in its possession.

The strain, however, was great. Confidence was again impaired. Many accounts were withdrawn by customers. We were compelled to ease our load by selling accumulated stocks at a loss. The price of Ely Central and other Scheftels promotions dropped. The decline was assisted by general weakness in other Curb stocks.

Peculiarly enough, at the time when the market for Ely Central shares was lowest, during the latter part of September, fourteen months after the Scheftels company had taken hold of the proposition, mine reports were most favorable. Underground development work and churn drilling had set at rest for all time the question of whether or not mineralized porphyry underlies the rhyolite cap or flow extending eastward through Ely Central ground from the steam shovel pit of the Nevada Consolidated. Upward of $240,000 had been paid out for administration, mine equipment and for miners' wages to make this demonstration.

The Scheftels company was now informed that the Nevada Consolidated was actually meditating a trespass on the Juniper claim of the Ely Central Co., in order to secure an outlet from the lower levels of its giant steam shovel pit. Warning in writing had already been served on the Nevada Consolidated officials against such a course. On September 25th attorneys of the Ely Central Copper Co. secured from a Nevada court an order restraining the Nevada Consolidated from proceeding with this trespass and citing it to show cause why it should not cease to trespass on other Ely Central ground.

The attorney telegraphed to New York that a bond was required before the injunction could be made operative. On September 27th and 28th telegrams were exchanged between the Ely Central offices in New York and the Nevada attorneys of the company at Reno as to providing sureties for the bond.

The sureties never qualified. A catastrophe befell us and brought to an earthquake finish the house of B. H. Scheftels & Co. and all of its ambitious plans.

The constant turmoil in which the House of Scheftels had found itself, from the day the *Engineering & Mining Journal's* attack appeared, had made it impossible for the Scheftels company to hold the markets for Ely Central and Rawhide Coalition. Impairment of credit, money stringency and a general declining market were partly responsible. But there was another important factor. Because of the time limit of its options, the Scheftels company was forced, from time to time, to throw stock on the market at prices which showed an actual loss.

It had one market winner which showed customers and the corporation itself a large profit, namely, Jumbo Extension. I held an option on approximately 450,000 shares of this stock at an average price of 35 cents, which I had turned over to the corporation. The market advanced to 70. Following the tactics employed in Ely Central at the outset of that deal, the Scheftels corporation had urged all its customers to buy Jumbo Extension at the very moment when I was negotiating for the option in Goldfield, with the result that purchases were made in the open market by readers of our market literature at from 25 cents up, with accompanying profit making.

As the price soared, a short interest of 150,000 shares of Jumbo Extension had developed among brokers in San Francisco and New York, and it was very apparent from the demands of stock for borrowing purposes that it would be impossible for the short interests to cover excepting upon our terms. The Scheftels company was making ready for a squeeze of the shorts such as had not been administered before in the history of the Curb. At the very moment of victory, however, when we were making ready to execute a magnificent market coup in Jumbo Extension in the markets of San Francisco and New York, we were plunged without warning to complete ruin.

THE RAID ON B. H. SCHEFTELS & CO.

THE DESTRUCTION OF the Scheftels structure was consummated on the 29th day of September 1910. I was standing on the front stoop of the Scheftels offices, watching the markets for the Scheftels specialties. A broker with San Francisco connections made me a bid of 68 cents for 10,000 shares of Jumbo Extension. I promptly refused. At that very moment my attention was called to the violent slamming of a door behind me. Turning to a Scheftels employee who was standing by my side, I learned that a number of strangers had filed into the customers' room without attracting any particular attention. I tried to get in. The door was locked. Undoubtedly something serious was transpiring. I walked a full block through the hallway to the New Street entrance of the building where the offices of the *Mining Financial News* adjoined those of the Scheftels corporation. I tried the door there with similar result. It was locked against me.

That settled it. I concluded that the ax had fallen.

The shock of realization that our offices were being raided by the government did not for a moment throw me off my balance or put fear in my heart, nor did the sense of the outrage affect me at the moment. There was but one sickening thought—the ruin of the edifice I and my associates had labored day and night for so many months to build and the fate of our customers who had invested their money in the companies we had promoted.

In three seconds I was on my way to the place where I thought succor could be found—the offices of the Scheftels attorneys. I walked across the street to the New Street entrance of a building that extends from Broadway to New Street, ambled across to the Broadway side, jumped on a surface car, rode three blocks to Broadway and Cedar Street, jumped into an elevator, and in a few minutes entered the offices of House, Grossman & Vorhaus.

"Go over to the Scheftels offices," I said, "and be quick! I think we are being raided."

In a moment two members of the law firm were on their way. Within ten minutes after the raiders had entered the offices the lawyers were on the spot. They were denied admittance and had to content themselves with waiting outside the door until the prisoners were taken out.

The moment the lawyers left their offices I began to use the 'phones to provide for the release on bail of the men arrested. I found it necessary to go in person and so I left the lawyers' offices and walked down Broadway. My attention was attracted by the clanging of the bell of the police patrol wagon. As it wheeled past me on the run I could see my associates huddled together in the Black Maria on their way to the bastile.

For the moment, I lost full sense of the gravity of what was transpiring and was overcome by a feeling of joy that I had been spared that ignominy. That self-felicitating slant of an intensely serious situation passed. My associates were in trouble and it was up to me to help them. I was at large and I knew that I could do more for my friends and myself by not immediately surrendering.

I returned to the lawyer's office, where I remained. All this time the thought never entered my mind that we were in any sense guilty of any intent to defraud anyone, or that we had committed any offense against law or the rules of fair conduct. The one consuming and controlling idea in my mind was that somebody had put one over on us and that it was up to me to organize for defense against the abominable outrage.

What transpired behind the closed doors while the Scheftels lawyers were attempting to gain an entrance for the instruction of the corporation, its officers and employees as to their rights, beggars description. Gentle reader, you would not conceive the reality to be possible. Armed with a warrant which conferred upon him the right to arrest, seize, search and confiscate, the special agent of the Department of Justice had secured from the local police headquarters a detail of fifteen heavily armed plain clothes men.

Once inside the Scheftels establishment, the doors were locked and egress barred. The main body of invaders then took possession of the front offices, while others searched through the back rooms and boisterously commanded everyone to remain where they were until given permission to depart. The establishment was under seizure, every foot of it, and every person found within its doors was held prisoner. The special agent took pains to impress upon everybody within hearing that he was in supreme command. Leaving police guards in the front room, he stalked into the telegraph cage where two or three operators were sitting at tables.

Pressing the muzzle of a revolver into the face of Chief Operator Walter Campbell—a quiet and inoffensive man—the special agent commanded:

"Cut off that connection!"

Mr. Campbell didn't at first see the gun because it was pointed at his blind eye. When he got his first peep he concluded that a maniac had invaded his sanctum and he almost expired with apoplexy on the spot.

Returning to the front offices, the agent entered the cashier's cage and took possession of the company's pouch containing its securities.

He gave no receipt to any responsible employee of the Scheftels company for anything. When Mr. Stone, one of the cashiers, suggested to him that he was there to safeguard the securities, he thundered, "Come out of there!"

"What authority have you for this?" demanded Mr. Stone. The agent thereupon showed his badge.

A moment later one of the deputies pried open the cash drawer. The special agent was at his elbow.

"Oh, look what's here!" cried the deputy.

Thereupon the agent of the Department of Justice impounded the contents of the cash drawer, without counting the cash, checks, money orders, etc., or giving any member of our firm a receipt for them.

Turning to the Scheftels officers and employees who had been placed under arrest, he ordered them removed from the room.

It was about as raw a performance as was ever witnessed in a peaceful brokerage firm's banking room.

Bookkeepers were ordered to close up books. United States mail in the office was impounded, including mail that had been received in the office for delivery to others.

The Scheftels employees were commanded to stand in their places with arms folded. The desperadoes among them—those for whom a warrant had been issued and who had been jerked out and huddled together in the outer room—were then searched for deadly weapons. One penknife and the stub of a lead pencil were found on their persons. The deadly knife was hardly sharp enough to serve the purpose of nail manicuring. Not one of the men under arrest would have known how to use a revolver if it had been placed in his hands.

The men taken into custody were: Mr. Scheftels, aged 54, quiet and inoffensive, rounding out an honorable business career without a blemish of any kind on his character or standing; Charles F. Belser, one of the cashiers of the corporation and a 32nd degree Mason, who never before in his life had been so much as charged with the violation of the spirit of a minor ordinance; Charles B. Stone, aged 60, another cashier whose sons and sons-in-law had served their country in the army and who, himself, was as peaceful as a class leader in a Sunday school; John Delaney, Clarence McCormick, William T. Seagraves and George Sullivan, clerks of the establishment, who were as likely to offer resistance that would require gun play to combat as were a quartette of psalm-singing children.

Mr. Scheftels protested in a dignified and self-respecting way against the brutal demonstration. He asked to see the authority for the raid. This was refused until after he and the other desperate characters were collected in another room. His demand to see the officers' warrant was met by a vulgar exclamation from the special agent, to the effect that, "If you don't shut up, we will put the irons on you! If you are looking for any trouble, you ———— ———— stiff, you will get what you are looking for!"

The absurdity of the armed invasion appealed to everybody but the ringleader of the raiders. It was a ludicrous situation from a service viewpoint. There had been no time up to the moment of the raid when a single man armed with proper authority could not have accomplished with decency and in good order everything and more than was done by the rough house and brutal invasion of the armed band.

Private papers were grabbed and bundles of certificates of stock, packages of money, checks, receipts and everything that came in sight were carried away. No complete record was made at the time of the raid of the documents and other valuables seized. The temporary receiver for B. H. Scheftels & Co., before his discharge later on, was able to gather together and take an accounting of part of the seized assets of the corporation, but I have no doubt that many thousands of dollars worth of securities and money were hopelessly lost.

When the wreck was complete the prisoners were driven like malefactors out of the front entrance, down the steps and loaded into the Black Maria. Five thousand people witnessed the act. The prisoners pleaded in vain to be allowed to pay for taxicabs to convey them before the United States Commissioner. They urged that as yet they had no hearing and were innocent in the eyes of the law, and until convicted of some offense they were entitled to decent treatment. This request was refused. The delay in the start to the federal building was just long enough to give the dense crowd that had filled the block time to insult the victims of the atrocity to the fullest extent. Friends of the arrested men boiled over with indignation and several fights occurred. Men were knocked down, trampled upon and the clothing torn from their backs in the desperate mêlée. The scene was disgraceful.

An army of newspaper reporters, attended by a camera brigade, were on the spot and snapshotted the prisoners as they entered the Black Maria. With bells clanging and whips lashing, a start was made up Broad Street to Wall. Then the vehicle turned up Broadway and ding-donged on to the federal building. There the men were arraigned. Bail aggregating $55,000 was demanded. Later several of the men taken into custody in this brutal manner were not even indicted.

Called upon to identify the prisoners, the special agent of the Department of Justice was unable to point out any of them except Mr. Scheftels. A stenographer in the employ of the corporation was forced to single them out.

The warrant proved to have been sworn to by the special agent and had been granted on his affidavit that the corporation had committed crimes against some few of its customers. Two of them have already been mentioned in this article as Slack and Szymanski, whose statements had been furnished to the attorneys of the *Engineer & Mining Journal*.

From the Court House to the Tombs, the Scheftels desperadoes, in shackles, were escorted up Broadway. Later in the day when bail was ready and the prisoners were sent for, they were handcuffed again and marched in parade up streets and down avenues of the densest section of New York City.

I had worked all afternoon in the lawyers' offices with one object in view, namely, the securing of bail for the imprisoned men. I succeeded. I now got busy with my own bail, the court having fixed it in advance at $15,000. In the morning I walked from my lawyers' offices to the Post Office building and surrendered myself, being immediately released on surety which was waiting in the office of the United States Commissioner. As I left the building I recognized scores of Scheftels customers. Several grasped my hand.

My indignation grew as the circumstances came up under review and I had time to connect and collate the facts. Gradually the whole truth revealed itself. I can relate only part of it. The full, detailed story would extend itself into a volume, and the space here at my command is limited. I learned that from the moment the special agent had been put on the scent with permission to put us out of business he had never slackened his effort to turn the trick.

His efforts attracted the attention of sundry newspaper editors with Wall Street affiliations and also enemies generally who hastened to coöperate with him. His office as a special agent of the Department of Justice gave to his statements weight which would not have been given to them had he as an individual sponsored the charges. His official position imparted exaggerated importance to his statements in the eyes of newspaper men, and, after the raid, to the public.

A person whom we shall characterize as the Tool now appears on the scene with alleged information which he placed at the service of the special agent to back him up before the assistant United States attorneys in New York with testimony since recanted over the signature of the false witness.

Previously I called attention to some of the atrociously false statements that were published on the day following the raid. I gave only an inkling. The newspapers declared that Ely Central had cost the Scheftels company five cents per share, that the capital stock was over issued, and that the property was worthless. Jumbo Extension, which has since distributed $95,000 in dividends to its stockholders, has still a treasury reserve of $100,000 and is selling today in the markets at a share valuation of about one-quarter of a million dollars for the property, was also described as a fake stock. Rawhide Coalition, which has produced upward of $400,000 in bullion, and which is to-day recognized as one of the substantial gold mines of the Far West, was labeled plain junk. Bovard, which represented an investment of nearly $100,000 for property account and mine development and which had been promoted at 10 cents per share on representations that it was a prospect, was stated to be a raw steal.

The Scheftels corporation was said to have got away with millions of dollars by selling fake mining stocks. It was also stated that I had profited to the extent of millions for my personal account. The Scheftels mailing list was described as a regulation sucker list, notwithstanding the fact that the principal names that were on it were stockholders in Guggenheim companies.

The ringleaders were pictured as myself—"a man with an awful past"—and the "notorious character," "Red Letter" Sullivan. Mr. Sullivan was styled as the facile letter writer who had addressed the suckers and hypnotized them, principally widows and orphans, to withdraw their money from the savings banks and send it to the Scheftels sharks. "Red Letter" Sullivan was also referred to as a man with a past.

The true facts regarding Mr. Sullivan's connection with the Scheftels company were these: A few months before, he had applied for a position. He was then employed as a manager of a Boston stock brokerage office. He was awarded the job of time clerk in the stenographers' department.

His job, while employed with the Scheftels company, was to see that the stenographers reported on time, did their work properly and were not paid for any services they did not render. He had little or nothing whatever to do with the correspondence department. He never dictated any answers to letters received by the Scheftels company. Never was he employed in an executive capacity by the Scheftels company. We knew little or nothing of the "red letter" title with which he had been decorated. The first we learned of it was in the newspapers after the raid. Investigation revealed that ten years before, while a broker in Chicago, he had issued a weekly market letter which was printed on red paper.

I have thus far not given space to one of the greatest wrongs connected with this disgraceful proceeding—the wrong and damage inflicted upon a multitude of helpless stockholders. While the special agent of the Department of Justice and his armed followers were wrecking the Scheftels offices and terrorizing the place, the Scheftels group of mining stocks was being savagely raided on the Curb and enormous losses were inflicted on the public. Thousands of margin accounts were wiped out in less time than it takes to tell of the massacre. Declines in Ely Central, Jumbo Extension, Rawhide Coalition and Bovard Consolidated exceeded $2,000,000. This loss was distributed among approximately fourteen thousand shareholders of record and as many more not of record.

This large army of innocent shareholders was helpless. From such species of confiscation the law affords no relief or recourse, except actual acquittal of the arrested persons, in whom lies the confiding investor's only chance for the market rehabilitation of his securities.

A TOOL'S CONFESSION

THE SIGNED CONFESSION of the Tool of the special agent, who appeared before Assistant Attorneys Dorr and Smith at the United States Attorney's office in New York, which says he gave false testimony, and the voluntary statement of John J. Roach, a stock broker who was employed by the now defunct firm of Frederick Simmonds, regarding the relations between the special agent and that firm, while special agent of the government, reveal the weak foundations of the Government's charges.

The Tool, prior to the raid, had been in the Scheftels employ. For a few months he had been a traveling business getter for the firm. Then he was discharged. He associated himself with Frederick Simmonds, a member of the Consolidated Stock Exchange. Mr. Simmonds was badly in debt. The Tool had no money. The agent, when he was trying to get the United States Attorney's office in New York to agree that the information collected was sufficient to warrant a raid, prevailed upon the Tool to appear before the assistant attorneys and give testimony.

In this story the chief value of the Tool himself is that he has no value. He made his statement against us to Mr. Dorr, assistant U. S. district attorney. Then he gave me a statement, signed in the presence of witnesses, recanting the statements made to Mr. Dorr. To this he later added a written postscript enforcing his recantation. Then he re-recanted and said that a large part of his first recantation, signed by him and initialed by him on each page with his initials was false. The reader is left to judge just which one of the Tool's three positions is the one in which he tells the truth. It is obvious that he must be lying in the two others, and it is not impossible that he may be lying in all three—except that some of the stuff in his first recantation, which he later denies in his second, has been verified from other sources.

Here is the main point to bear in mind concerning the Tool,—the sovereign power of seizure, search and confiscation brought into play by our great government without due process of law, was based in part on the flimsy testimony of such a person. Thousands of investors suffered from the blow, as well as myself and associates.

It would appear from the Roach statement that he was largely instrumental in bringing about the crisis that resulted in the suspension of the Simmonds firm and in the disclosures of the special agent's relations therewith. These facts have become, in most instances, matters of public record. They came out during the hearings before the receiver for the bankrupt concern. It was found that the liabilities of the busted firm were $85,000 and the assets 100 shares of cheap mining stocks and between $1,500 and $2,000 in cash. It was at this conjunction that the special agent was allowed to resign from the Department of Justice. The Tool he had foolishly used had proved to be a two-edged one. The agent had been "hoist by his own petard."

THE GUGGENHEIMS

PROBABLY THE MOST surprised branch of the Government at the time of the Scheftels raid was the Post Office Department. The crime charged was misuse of the mails. Why, if the Scheftels aggregation were guilty, didn't the Post Office Department do the raiding? Why didn't it issue a fraud order? The Scheftels company has since been declared solvent by the courts and the temporary receiver discharged. To this day no fraud order has been issued. Only a short period before the raid, a presentation on the part of the Post Office Department of all of the evidence in the case had been met with a decision that there was no ground for action.

That the Guggenheim interests did not fail to take advantage of the plight of the House of Scheftels immediately after the raid finds conclusive proof in the transpirations in the Ely mining camp. Soon after the special agent descended on the Scheftels offices, an application in Ely was made for a receiver to take charge of the assets of the Ely Central Copper Co. The attorneys making the application were Chandler & Quale, attorneys for the Nevada Consolidated Copper Co., a Guggenheim enterprise. When the court appointed a receiver he named this firm as attorneys for the receiver. Attorney J. M. Lockhart for the Ely Central made a protest that these lawyers, because of their connection with the Nevada Consolidated were not the proper persons to protect the interests of the now defenseless Ely Central stockholders. Then the court appointed another attorney, named Boreman.

Shortly after the receiver was appointed, he applied to the courts for permission to sell to the Nevada Consolidated for $30,000, which represented the entire cash indebtedness so far as the receiver knew, the surface rights to a large acreage of Ely Central and the rights through Juniper Canyon. This, if accomplished, would have given to the Nevada Consolidated a railroad right of way that would have solved the problem confronting it of the transportation of the ores from the lower levels of the steam shovel pit. Without such an outlet these ores could not have been handled without great expense and much difficulty. The benefits that would have accrued to Nevada Consolidated were almost incalculable. At the same time, such action would effectually cut the Ely Central property into two parts. According to the petition it was stipulated that in selling the surface rights the Ely Central should cede to the Nevada Consolidated practical ownership, because it was specified that Ely Central could not interfere in its mining operations with any rights granted. Attorney Lockhart of Ely Central fought the receiver and his attorneys and won a victory. The Ely Central property was saved intact for the stock holders.

Later, an application was made to the court to sell the entire property of the Ely Central for $150,000. This was believed to be in the interests of the Nevada Consolidated. In answer, a petition was filed to discharge the receiver on the ground that the court originally appointing him had no jurisdiction. The court finally decided that it was without jurisdiction, because neither fraud nor incompetency had been proved, and the property had not been abandoned. The receiver was discharged.

What has been the attitude of the Department of Justice since the raid was made? Since the raid the government has spent several hundred thousand dollars to disclose sufficient evidence from the books to make a case of any kind. One stand after another has been taken only to be abandoned after exhaustive research for evidence to sustain the original excessive pretenses. Grand Jury after Grand Jury has thrashed over masses of evidence presented them. Armies of accountants have worked day and night for weeks and months in an effort to substantiate the action of the authorities who were led into the commission of a grave wrong.

The charge that the Scheftels corporation sold fake mining stocks has fallen to the ground. Government examinations of the properties have revealed them to be all that they were cracked up to be. Careful and industrious reading of the mass of market literature sent through the mails by the Scheftels corporation has failed to disclose deliberate misrepresentation regarding the potentialities of any of the mining properties.

The Scheftels corporation transacted considerable margin business with its customers in the stocks which it sponsored—Ely Central, Jumbo Extension, Rawhide Coalition and Bovard Consolidated. If the Scheftels corporation was run by rascals wouldn't they have been tempted frequently to throw their weight on top of the market and endeavor to break the price of stocks to wipe out the margin traders? Did the government find any evidence of this in the books? No. It found evidence—overwhelming and cumulative—that on nearly all occasions the Scheftels corporation actually exhausted its every resource to support the market in its stocks and hold up the price in the interests of stockholders. Evidence was also found in quantity that the Scheftels company discouraged the practice of margin trading.

The superseding indictment handed down by the Grand Jury late in August 1911, eleven months after the raid, eliminated the charge of mine misrepresentation regarding the Scheftels promotions and reduced it practically to one of charging commissions and interest without earning them.

Not less than 85 per cent of the total brokerage transactions of the Scheftels corporation were in their own stocks, and at nearly all times in the Scheftels history it had on hand, put up on loans or in banks under option, anywhere from three million to seven million shares of these securities. It actually bought, sold and *delivered* in this period over fifteen million shares of stock!

As already stated, the Scheftels corporation made it a practice to sell stocks on the general list as an insurance against declines in the market which might carry down the price of its own securities, and this, in the finality, was what the government, after the expense of hundreds of thousands of dollars and the employment of the wisest of counsel, was compelled to tie to in order to justify in the eyes of the great American public the use of the rare power of seizure, search and arrest and of its denial of a prayer for a hearing to the victims which was made before the arbitrary power was used.

12. The Lesson of It All

WHAT IS THE LESSON of my experience—the big broad lesson for the American citizen? This is it: Don't speculate in Wall Street. You haven't got a chance. The cards are stacked by the big fellows and you can win only when they allow you to.

The information that is permitted to reach you as to market probabilities through the financial columns of the daily newspapers is, as a rule, poisoned at its fountain. It has for its major purpose your financial undoing. Few financial writers dare to tell the whole truth—even on the rare occasions when they are able to learn it. Most of them are, indeed, subsidized to suppress the truth and to accelerate public opinion in the channels that mean money in the pockets of the securities sellers. As for the literature of stock brokers it is generally even more misleading. Few brokers ever dare to tell the whole truth for fear of embittering the interests and being hounded into bankruptcy and worse.

As for myself, what excuse have I had for catering to the gambling instinct? This is it: I thought the promoter and the public could both win. I now know that this happens only rarely. As the game is now generally played by the big fellows, the public hasn't got a chance.

I have not got a dollar. Who profited? The answer is: if anybody, the aggregate. The world has been the gainer. It is richer for the gold, the silver, the copper, and other indestructible metals that have been brought to the surface, as a result of this endeavor, and added to the wealth of the nation.

But for the gambling instinct and the promoter who caters to it, the treasure stores of Nature might remain undisturbed and fallow and the world's development forces lie limp and impotent.

THE END

Index

Adventure Magazine, 16
Aldrich, Sherwood, mine owner, 74
Alexander lease, Rawhide, 215
Allen, Wing B., newspaperman, 161
Amethyst mine, Bullfrog, 74, 81
Anaconda Copper Co., 307
April Fool claims, Manhattan, 87
Arkell, W. J., newspaperman, 53-60, 81, 84
Atlanta mine, Goldfield, 71, 116

Baker, Ray, 228
Balloon Hill, Rawhide, 214-215, 234
Bank of Nevada, Reno, 159
Bartlett, George A., Congressman, 141, 206
Baruch, B. M., financier, 189-190, 196-198, 291
Bayley, William, mine director, 140
Beatty, A. Chester, mining engineer, 178, 181-183
Belcher Extension mine, Weepah, 19
Bingham-Galena Mining Co., Utah, 18
Blackwell's Island prison, 17
Bleakmore, Lou, stockbroker, 94
Blue Bull mine, Goldfield, 129, 130
Boas, Nat, stockbroker, 270
Bond, W. F. "Billy," mine promoter, 86
Booth mine, Goldfield, 129, 130
Boreman, ———, attorney, 322
Bovard Consolidated mine, 280, 302, 309, 319, 320, 323

Boyd, D. B., Washoe Co. Treasurer, 12, 188
Bragdon, H. T., mine director, 140
Brock, John W., mine director, 138
Broken Hills, Nev., 18
Broken Hills Silver Corp., 18
Bryan, Ben, stockbroker, 169
Buckskin, Nev., 20
Buckskin National mine, Humboldt Co., 20
Bullfrog, Nev., 10, 11, 71-84, 221, 224
Bullfrog Extension mine, 81
Bullfrog Gibralter mine, 107
Bullfrog Homestake mine, 80
Bullfrog National Bank mine, 74
Bullfrog Rush Mining Co., 11, 109, 111-114, 156
Bullfrog Steinway mine, 80

Camp, Sol, mining engineer, 87
Campbell, John Douglas "Jack," mine manager, 97-99, 103-104, 107, 109, 138, 141
Chandler & Quale, attorneys, 322
Clark, H. H., mine director, 140
Clark, W. H., mine promoter, 107
Clay, Christopher C., attorney, 163
Clipper claim, Ely Central mine, 305
Columbia Emerald Development Corp., 19
Columbia Mountain mine, Goldfield, 129, 130, 150

Combination Fraction mine, Goldfield, 114, 124, 150
Combination mine, Goldfield, 11, 86, 104, 124
Conqueror mine, Goldfield, 130
Consolidated Hill, Rawhide, 214-215
Constant, C. L., Co., mining engineers, 304-305, 307
Constant, S. C., mining engineer, 15
Converse, E. C., banker, 179, 186, 291
Cook, John S., banker, 111
Cook, John S. & Co., bank, 109, 126-127, 131, 151, 159, 164, 167, 197, 205-206
Copper Flat ore-body, Ely, 294
Coulson, Duncan, attorney, 179
Crackerjack mine, Goldfield, 129, 130
Cumberland-Ely mine, Ely, 290

Daily America, 49-50
Daisy mine, Goldfield, 71
De Lamar, Capt. Joseph R., mine owner, 178-179, 186
Denver Rush Extension mine, Bullfrog, 81
Dexter mine, Manhattan, 85, 91
Diamondfield, Nev., 15, 98, 150
Dick Bland fraction, Goldfield, 15
Dickerson, Denver S., Lt. Governor, 157
Dodge, Edmund R., attorney, 311
Dorr, ———, Ass't. U.S. Atty., 320, 321
Dunham, Sam C., editor, 77-78, 80
Dunnaway, J. F., manager NC&O RR., 12, 188

Eagle's Nest Fairview Mining Co., 109, 111, 116, 118, 156, 177
Eagle's Nest mine, Fairview, 108-109
Earle, E. P., geologist, 179, 186, 291
Eclipse mine, Bullfrog, 81
Edwards, Dan, mine promoter, 13, 199
Edwards, Tom, merchant, 75
Elliott, C. H., mine promoter, 62, 71, 74, 86, 87-91
Ely, Nev., 294, 303-304, 322
Ely Central Copper Co., 15, 244, 246-248, 280, 291, 294-311, 313-314, 319, 323

Engineering & Mining Journal, 17-19, 298-300, 303-304, 306-309, 313, 318
Epstine, Henry E., stockbroker, 141
Ethel mine, Goldfield, 130
Eureka workings, Ely Central mine, 305

Fairview, Nev., 108-109
Fairview Eagle mine, 109
Fairview Hailstone Mining Co., 118, 156
Farish, Col. William A., mining engineer, 296, 304, 305
Farmers & Merchants National Bank, Reno, 159
Fidelity Finance & Funding Co., Reno, 18
Financial Watchtower, The, 20-21
First National Bank of Goldfield, 87
Florence mine, Goldfield, 98
Four Aces mine, Bullfrog, 81
Frances-Mohawk Mining & Leasing Co., Goldfield, 140
Fuller-McDonald lease, Goldfield, 98
Funston, ———, Brig. Gen., U.S. Army, 206

Gans, Joe, prize fighter, 118-119, 122-123
Gans-Nelson prize fight, Goldfield, 118-119, 122-123
Gates, Charles G., 234
Gates, John W., stockbroker, 138, 234
Gibralter mine, Bullfrog, 80, 81
Gillies, Don, mining engineer, 76, 138
Giroux mine, Ely, 294, 304
Glyn, Elinor, author, 14, 227-228, 230, 232
Gold Bar mine, Bullfrog, 81
Golden, Frank, banker, 93
Golden Hotel, Reno, 126
Golden Scepter mine, Bullfrog, 81
Goldfield, Nev., 10-13, 15, 53-71, 81-87, 90-107, 114-134, 147-178, 187-209, 245
Goldfield Bank & Trust Co., 60, 61
Goldfield Consolidated mine, 13, 15, 64, 67, 86, 108, 126, 129-134, 140, 147, 150-151, 154-155, 157, 159, 163, 167, 178, 189-190, 195-199, 201-202, 205-206, 208, 240, 245

Index

Goldfield Daily Tribune, 15-17
Goldfield Daisy mine, 150
Goldfield Laguna mine, 67, 91, 127, 129, 130
Goldfield Mine Owners' Assoc., 205
Goldfield Mining Co., 71, 127, 129, 130, 140
Goldfield News, 12, 14-16, 20-21, 61, 175
Goldfield Stock Exchange, 116, 124, 127, 151, 154
Goldfield-Tonopah Advertising Agency, 10, 61, 161
Gold Wedge fraction, Goldfield, 15
Goodrich, Edna, entertainer, 13
Goodwin, Nat C., comedian, 13-14, 199, 235, 267, 269, 273, 298-299, 308
Goodwin, Nat C., & Co., 13-14, 199, 201, 215, 218, 245, 253, 255-256, 263-267, 270, 273, 278
Grannan, Riley, gambler, 235, 238
Grant, Peter, co-owner of Palace Club, 10, 91, 111-112, 170-171
Great Bend mine, Goldfield, 71, 150
Greenwater, Cal., 11, 80; mines at, 135-146
Grutt, Eugene, mine owner, 14
Grutt Hill, Rawhide, 214-215, 234
Guggenheim brothers, financiers, 178, 180-183, 290, 291, 295, 307, 322

Hall, Robert C., stockbroker, 77-79
Hammond, John Hays, mining engineer, 180-183, 291
Hayden, Stone & Co., bankers, 255, 265, 268, 276
Hayes and Monnette, mine owners, 107, 116
Haynes-Monnette lease, Goldfield, 66, 104, 107, 157
Hedrick, Harry, reporter, 116, 239, 277
Herzig, Charles S., 15, 17, 304-307
Herzig, Jacob Simon, *see* George Graham Rice
Hibernia mine, Goldfield, 130
Hoffman, Ed, mine superintendent, 225
Hooligan Hill, Rawhide, 215, 232-233

Hopper, James, author, 116, 224
Horton, Frank, mine promoter, 150
Howell, Eugene, banker, 140
Hungerford, Edward, author, 247-248

Idaho Copper Company, 19
Idaho Copper Corp., 19
Indian Camp mine, Manhattan, 98-99, 101-102, 111, 116, 156, 177
Ingalls, Theodore, Chief Inspector, 311

Jones, January, mine owner, 62, 91
Jordan, Joseph S., editor, 239
Jumbo Extension mine, Goldfield, 11, 15, 71, 114, 245, 280, 311, 313-314, 319-320, 323
Jumbo Mining Co., Goldfield, 15, 71, 98, 116, 124, 126, 127, 129, 130-131
Jumping Jack Manhattan Mining Co., 92-95, 98-99, 111, 116, 156, 175
Jumping Jack mine, Manhattan, 10, 86, 91-92, 224
Juniper Canyon, Ely, 322
Juniper claim, Ely Central mine, 313

Kearns leases, Rawhide, 214
Keith, Frank, mine manager, 139
Kendall, Zeb, mine owner, 91
Kendall mine, Goldfield, 129, 130
Kernick lease, Goldfield, 98
Kewanas mine, Goldfield, 71, 130, 150
Keystone Copper Mining Co., 290
King, E. W., mine manager, 14, 256
Kirby, John A., mine director, 140
Knickerbocker, H. W., promoter, 238
Knickerbocker Trust Co., bankers, 201
Krumb, Henry, mining engineer, 77
Kunze, Arthur, mine owner, 140

Lindsey, J. L., banker, 141, 169-171
Lockhart, J. M., attorney, 322
Lockhart, Thomas G., mine owner, 14, 77
Loftus-Sweeney lease, Goldfield, 124
Logan & Bryan, wire service, 168-169, 269, 270
Lou Dillon Goldfield Mining Co., Goldfield, 124-125, 156, 160, 177

Lou Dillion mine, Goldfield, 11
Luce, Ben, mine owner, 108-109
Lyman, Dr. J. Grant, mine promoter, 109, 111-114

Macdonald, Malcomb, mining engineer, 56-57, 74-77, 135, 138-140
Mackenzie, Donald, mine promoter, 140, 171
Manhattan, Nev., 10, 11, 86-105, 224
Manhattan Buffalo mine, 87, 90
Manhattan Combination mine, 87, 90
Manice, E. A., & Co., stockbrokers, 141, 144
Maxim & Gay, tipster service, 9, 30-50, 163
May Queen mine, Goldfield, 129, 130
McCafferty, ———, police inspector, 307-308
McCornick, Henry, banker, 99, 103
McCornick & Co., bankers, 99
McDaniel, ———, mine superintendent, 188
McKane, John, mine owner, 77, 98
Menardi, Capt. J. B., mine owner, 126
Meyer, Eugene, Jr., 180
Miller, Al, mine operator, 232-234
Miller, Charles R., pres. of T&GRR., 138
Miller, Warren A., mining man, 13, 199
Miller lease, Rawhide, 233
Milltown mine, Goldfield, 129, 130, 150
Mims-Sutro Co., Goldfield, 62
Miner's Hospital, Goldfield, 126
Mining & Metallurgy, 22
Mining Financial News, 14, 275-279, 283, 287, 289-290, 294, 295, 303, 311, 314
Minzesheimer, Charles S., & Co., 57, 138, 146
Mitchell, J. F., mining engineer, 125, 157
Mizpah mine, Tonopah, 79
Mocha Café, Goldfield, 70
Mohawk Mining Co., Goldfield, 11, 66, 70-71, 105, 114, 116, 124, 126, 127, 129, 130-131, 150, 157

Monarch Shaft, Ely Central mine, 297, 305
Monnell, Ambrose, mine owner, 179
Montana-Tonopah Mining Co., 56
Montezuma Club, Goldfield, 155
Montgomery, Bob, mine owner, 75
Montgomery, Hugh, publisher, 190
Montgomery Mountain mine, Bullfrog, 81
Montgomery-Shoshone mine, Bullfrog, 74-77, 80-81, 135, 139
Morgan, J. Pierpont, financier, 196, 240, 291
Murray, Lawrence O., Ass't. Sec. Dept. of Commerce, 206, 208
Murray lease, Rawhide, 215
"My Adventures With Your Money," published by *Adventure Magazine*, 16; published in book form, 17
Myers, Al, mine owner, 91

Neal, Charles B., Labor Commissioner, 206
Nelson, "Battling," prize fighter, 118-119, 122-123
Nevada Consolidated Copper Co., Ely, 290, 294, 296-297, 303, 305, 313, 322
Nevada First National Bank of Tonopah, 135
Nevada Hills mine, Fairview, 108-109, 140, 157
Nevada Mines News Bureau, 12, 189
Nevada Mining News, 12-13, 14, 133, 189-190, 192-193, 195-196, 198, 209, 270, 275
Nevada Smelters & Mines Co., 57
Nevada State Journal, 13, 19, 111, 196
Nevada-Utah Mining Co., 289-290
Nevada Wonder mine, Wonder, 188
Newhouse, Sam, mine operator, 98, 228, 230
Newlands, Francis B., U.S. Senator, 209, 311
News Letter, 241, 276
Nipissing, promotion of, 178-187

Index

Nixon, George S., U.S. Senator, 13, 14, 16, 74, 84, 91, 127, 129-134, 140, 150-151, 154, 157, 159-160, 163-164, 166-168, 173, 189-190, 192-193, 195-198, 201-202, 205-206, 208-209, 291
Nixon National Bank, Reno, 133, 159, 173, 192
Northern Saloon, Goldfield, 70
Nye & Ormsby County Bank, 93, 159, 165, 201-202, 205

O'Brien, Jimmy, newspaperman, 61
Oddie, C. M., mine director, 139
Oddie, Tasker L., Nevada Governor, 74, 84, 135, 139
Original Bullfrog mine, 81
Oro mine, Goldfield, 129, 130

Paine, Webber & Co., stockbrokers, 276
Palace Club, Goldfield, 10
Paloverde fraction, Goldfield, 15
Patrick, Elliott & Camp, Inc., mine promoters, 62, 87, 91, 94
Patrick, L. L., mine owner, 86
Peery, Henry, mine promoter, 107-109
Pell, S. H. P., & Co., stockbrokers, 179
Pheby brothers, mine owners, 295
Phoenix, ———, prospector, 125
Pierce, Richard T., mining engineer, 305
Polaris Mine, Tonopah, 79
Pollock, James A., & Co., stockbrokers, 170-171

Quartzite lease, Diamondfield, 98

Rawhide, Nev., 13, 211-256
Rawhide Coalition Mining Co., 14, 15, 215, 218, 225, 227, 232, 245, 253, 255-256, 263-270, 273, 277, 280, 288, 302, 311, 313, 319-320, 323
Ray Central Mining Co., 290
Red Hills mine, Goldfield, 129, 130
Red Top mine, Goldfield, 71, 114, 124, 126, 127, 129, 130-131
Reed, Walter Harvey, geologist, 19
Reilly lease, Goldfield, 98

Reno, Nev., 12
Reno *Gazette*, 192
Reveille, Nev., 57
Rhyolite, Nev., 71-84
Rice, George Graham (Jacob Simon Herzig), larceny conviction in New York, 9; forgery conviction in New York, 9; at Sing Sing, 9; adopts pseudonym, 9; operates Maxim & Gay tipster service, 9, 29-50; purchased *Daily America* newspaper, 49; arrives in Nevada, 10, 55; locates in Goldfield, 57; founds Goldfield-Tonopah Advertising Agency, 10, 61; promotes Bullfrog, 71-84; promotes Manhattan, 86-105; founds L. M. Sullivan Trust Co., 10, 95; promotes Fairview, 108-109; promotes Greenwater, 135-146; returns to New York (1907), 177; moves to Reno, 12, 187; promotes Wonder, 12, 188; becomes editor of *Nevada Mining News*, 13, 196; creates Nat C. Goodwin & Co., 13, 199; promotes Rawhide, 13-14, 211-256; joins B. H. Scheftels & Co. in New York, 14, 273; becomes editor of *Mining Financial News*, 277; divorced in Reno, 15; writes "My Adventures With Your Money," 16; convicted of mail fraud (1912), 17; convicted of grand larceny in New York, 18; operates Fidelity Finance & Funding Co. in Reno, 18; operates at Rocky Bar, Idaho, 19; invests in Weepah boom, 19; operates Columbia Emerald Development Corp., 19; convicted of mail fraud (1928), 19-20; in Atlanta federal prison, 20; separates from second wife, 20; released from prison, 20; begins publishing *The Financial Watchtower* in 1934, 20-21; in Milwaukee, 22
Rich Gulch Wonder Mining Co., Wonder, Nev., 12, 188, 199

Rickard, Tex, fight promoter, 14, 70, 118, 122, 253
Rickey, Thomas B., banker, 141, 174
Roach, John J., stockbroker, 320-321
Roosevelt, Theodore, U. S. President, 206, 208

Sandstorm mine, Goldfield, 129, 130
Scarborough, George, Special Agent of Dept. of Justice, 274
Scheftels, B. H., stockbroker, 277, 278, 307, 309-311, 317
Scheftels, B. H. & Co., stockbrokers, 14, 17, 244-246, 264, 270, 273-285, 287-290, 294-324
Scheftels Market Letter, 14, 275, 278, 283, 287, 289, 290, 295, 303, 308
Scheeline, M., banker, 202, 256
Scheeline Banking & Trust Co., Reno, 201-202, 256
Schwab, Charles M., financier, 75-80, 135, 138-139, 145, 291
Scott, W. H., stockbroker, 238
Scott & Amann, stockbrokers, 238
"Scotty," Death Valley, 227
Seyler-Humphrey Mining Co., Manhattan, 87, 90, 94
Shoshone mine, Tonopah, 79
Shoshone National Bank mine, Bullfrog, 80
Shumacher, C., & Co., bankers, 179
Siler, George, fight referee, 118-119, 122-123
Silver Pick Extension mine, Goldfield, 124-125, 156, 177
Silver Pick mine, Goldfield, 71, 114, 124, 150
Simmerone mine, Goldfield, 57, 60
Simmonds, Frederick, stockbrokerage, 320-321
Sing Sing prison, New York, 9
Slack, C. H., investor, 309, 318
Smith, ———, Ass't. U.S. Attorney, 320
Smith, Herbert Knox, Commissioner of Corporations, 206
Southern Pacific RR Co., 151, 154, 195

Sparks, John, Nevada Governor, 12, 109, 111, 113, 125-126, 156-157, 188, 201, 206, 208
State Bank & Trust, Goldfield, 107, 141, 159, 165, 169-172, 174, 201-202, 205
State Bank & Trust, Tonopah, 56
Stewart, William S., U.S. Senator, 221, 224
Stingaree Gulch, Rawhide, 230, 253
Stray Dog Manhattan mine, 86, 91, 98-99, 101-104, 111, 116, 156, 175, 177
Sullivan, L. M. "Larry," 10, 12, 16, 91-95, 97, 103-104, 107, 111, 114, 119, 122-123, 126, 127, 161, 171, 188, 196, 224, 319
Sullivan, L. M., Trust Co., 10-12, 16, 95, 98-99, 102, 104-105, 107-109, 111, 113, 116, 118-119, 122-126, 138, 141, 144, 145, 155-157, 160-161, 164-175, 177, 202, 208, 224
Sullivan & Rice, promotion firm, 188, 190, 196, 199
Sutro, Richard, banker, 141
Sutro Bros. & Co., bank, 141
Szymanski, D. J., investor, 309, 318

Taylor, "Billy," mine owner, 108-109
Taylor, Charles D., mine owner, 126, 127
Taylor, H. L., mine owner, 126
Teague, Merrill A., editor, 12, 189, 192-193, 196
Thomas, ———, Governor of Colorado, 161, 195
Thompson, R. M., mine owner, 179
Thompson, Towle & Co., stockbrokers, 241, 276
Thompson, W. B., mine promoter, 178-179, 180-183, 186-187, 241, 289-290, 295
Todd, J. Kennedy, & Co., financiers, 198
Tonopah, Nev., 55-57, 77
Tonopah & Goldfield RR Co., 138, 141
Tonopah Banking Corp., Tonopah, 140, 159
Tonopah Club, 71, 91, 154

Index

Tonopah Extension Mining Co., 77-79, 135, 138
Tonopah Home Consolidated mine, 84
Tonopah Home mine, 60, 81, 84
Tonopah Miner, 13, 78
Tonopah Mining Co., 75, 79, 138, 139, 144
Tripp, Alonzo, manager T&GRR, 141
Tramps Consolidated mine, Bullfrog, 74, 81, 109, 111
Turner, Oscar Adams, mine promoter, 75, 79, 144, 295
Tybo, Nev., 57

Union Securities Co., Goldfield, 111, 113
U.S. Mint, San Francisco, 173
Utah Copper Co., 290

Vermilyea, Edmonds & Stanley, law firm, 98
Vinegarroon fraction, Goldfield, 15

Wallace, "Black," 151, 154

Ward, Dr. M. R. financier, 79, 135, 138
Washoe County Bank, Reno, 159
Webb, U.S., Calif. Atty. Gen., 12, 188
Weed, Dr. Walter Harvey, geologist, 304, 306-307
Weepah, Nev., 19
Weir, J. C., stockbroker, 103-104, 113-114, 138, 270
Weir Brothers & Co., stockbrokers, 113-114
Wells-Fargo Express Co., 105
Western Federation of Miners, 126
Wharton, James, mine owner, 179
Wiley, Capt. W. Murdoch, mining engineer, 305-307
Wingfield, George, mine owner, 15, 16, 64, 71, 81, 91, 97, 108, 126-131, 140, 150-151, 154-155, 157, 159-160, 163-164, 166-168, 173, 190, 195, 197-198, 201, 205-206, 208, 291
Wonder, Nev., 12, 188
Woollacott, H. J., mine director, 140

Photo credits are as follows. The lower case "t" denotes illustration on top of page; "b" is on the bottom of a page. Betty Burgess and Leo Grutt, 8; Theron Fox, 4, 250t; Charles Gallagher, 292t; Frank Mitrani photo, 254.

Nevada Historical Society, 158, 184 (line sketch), 210; Hugh Shamberger collection (courtesy of Nevada Bureau of Mines and Geology), 134, 200, 203, 216t, 217b, 222, 226t; Hugh Shamberger collection (courtesy of Philip Earl), 120-121, 128; all others are from the Stanley W. Paher collection.

The text of this new edition of *My Adventures with Your Money* has been designed and entirely reset in Linotype Janson by Daniel Cronkhite at the Sagebrush Press in Morongo Valley, California.

❧

Inter-Collegiate Press in
Shawnee Mission, Kansas, prepared the plating,
conducted the printing by way of the lithographic/offset method, and oversaw binding operations through to completion.